ENERGY SCIENCE, ENGINEERING AND TECHNOLOGY

ALTERNATIVE WATER USE FOR POWER PLANT APPLICATIONS

ENERGY SCIENCE, ENGINEERING AND TECHNOLOGY

Additional books in this series can be found on Nova's website under the Series tab.

Additional E-books in this series can be found on Nova's website under the E-book tab.

WATER RESOURCE PLANNING, DEVELOPMENT AND MANAGEMENT

Additional books in this series can be found on Nova's website under the Series tab.

Additional E-books in this series can be found on Nova's website under the E-book tab.

ENERGY SCIENCE, ENGINEERING AND TECHNOLOGY

ALTERNATIVE WATER USE FOR POWER PLANT APPLICATIONS

ERIC N. GRADY
EDITOR

nova publishers
New York

For permission to use material from this book please contact us:
Telephone 631-231-7269; Fax 631-231-8175
Web Site: http://www.novapublishers.com

Library of Congress Cataloging-in-Publication Data

ISBN: 978-1-62808-986-8

Published by Nova Science Publishers, Inc. † New York

CONTENTS

PREFACE

A U.S. Department of Energy National Energy Technology Laboratory (DOE/NETL) study estimated that in 2005, total U.S. freshwater withdrawals for thermoelectric power generation amounted to approximately 146 billion gallons per day (BGD), while freshwater consumption was 3.7 BGD. Coal-fired power plants use water for cooling water systems, flue gas desulfurization (FGD) makeup, boiler makeup, ash handling, wastewater treatment, and general plant wash down. As the population continues to increase, electric and water demands are expected to grow. Thermoelectric power generation requires a large quantity of freshwater to support operations. In regions of the country with limited freshwater supplies, thermoelectric power generation may be competing with other uses for dwindling freshwater supplies. This book focuses on the research that has been done to date for the non-traditional sources of process and cooling water components. Studies in this research area focus on identifying potential new water sources, including brackish and saline water supplies and various domestic and industrial wastewaters. In particular, these studies focus on understanding the location, volumes, and quality of different non-traditional waters, as well as treatment technologies to reduce the scaling, biofouling, and corrosion potential of available impaired waters that would not otherwise be useful in a cooling water system or other processes within a thermoelectric plant.

Chapter 1 – A U.S. Department of Energy National Energy Technology Laboratory (DOE/NETL) study estimated that in 2005, total U.S. freshwater withdrawals for thermoelectric power generation amounted to approximately 146 billion gallons per day (BGD), while freshwater consumption was 3.7 BGD. Coal-fired power plants use water for cooling water systems, flue gas desulfurization (FGD) makeup, boiler makeup, ash handling, wastewater treatment, and general plant wash down. As the population continues to increase, electric and water demands are expected to grow.

Research examining the energy/water link in coal-based power plants is sponsored by DOE/NETL under the Existing Plants, Emissions and Capture (EPEC) Program. The research and development (R&D) program aims to minimize freshwater withdrawal and consumption without introducing negative impacts of poor water quality on plant operations. In order to develop cost-effective approaches to using lower-quality water sources, the Non-Traditional Sources of Process and Cooling Water component of the EPEC Program includes research focusing on potential new water sources, including brackish and saline water supplies and various domestic and industrial wastewaters. Studies in the treatment of municipal treated water/reclaimed water, produced waters from oil and gas wells, mine pool waters, produced waters from carbon dioxide (CO2) storage in saline formations, and ash pond basins focus on understanding the location, volumes, and quality of different non-traditional waters. This component also develops treatment technologies to reduce the scaling, biofouling, and corrosion potential of available impaired waters that would not otherwise be useful in cooling water systems or other processes within thermoelectric plants.

Reclaimed water (treated municipal wastewater) is widely available in communities throughout the United States in sufficient volumes and is reliable enough to supply power plant cooling water. Reclaimed waters are already being used in more than 50 U.S. power plants and are subject to Federal and state regulations in order to protect worker and public health. Researchers have developed small pilot-scale cooling towers for side-by-side evaluation of the use of impaired waters under different operating conditions. Standard chemical additives provided reasonable scaling and biofouling control. The corrosion rates were not affected by the corrosion inhibitors but were strongly influenced by the degree of scaling on the corrosion coupons. The effects of combinations of tertiary treatments (e.g., nitrification, filtration, carbon adsorption) and different chemical treatment regimens are under investigation.

Chaptyer 2 – Throughout the United States, many sizable saline formations may have the potential to hold carbon dioxide (CO_2), a capability that could be used to manage CO_2 emissions from power plants. This analysis examined the southeastern U.S. to determine where saline formations exist that could be used both to hold CO_2, as well as to provide saline waters for treatment to develop a coupled-use system to extract the water, treat it, and then use it to help with power plant cooling efforts. While this is a high-level study, the results indicate that limited saline formation/well data restricted the ability to adequately assess many regions of the southeastern U.S. In Texas, however, the Carrizo-Wilcox system met the size, location, depth, and other criteria to potentially host a coupled CO_2 sequestration system along with water extraction and treatment for use. Using the analytical framework applied in this analysis, other regions of the country and site-specific test cases may benefit from the insight provided by a coupled-use (or similar) system to increase the value of CO_2 storage to meet growing regional water stress on the surface.

In: Alternative Water Use for Power Plant Applications ISBN: 978-1-62808-986-8
Editor: Eric N. Grady

Chapter 1

USE OF NON-TRADITIONAL WATER FOR POWER PLANT APPLICATIONS: AN OVERVIEW OF DOE/NETL R&D EFFORTS*

Ronald Munson, James Murphy and Kenneth Walsh

LIST OF ABBREVIATIONS AND ACRONYMS

Acronym/Abbreviation	Definition
AMD	Acid Mine Drainage
ANL	Argonne National Laboratory
BC	Brine Concentrator
BGD	Billion Gallons Per Day
BOD	Biochemical Oxygen Demand
BOD5	Five-Day Biochemical Oxygen Demand
CBM	Coalbed Methane
CCS	Carbon Capture and Sequestration
CEC	California Energy Commission
CO2	Carbon Dioxide
COC	Cycles of Concentration
CWA	Clean Water Act
CWTS	Constructed Wetland Treatment Systems
CaCO3	Calcium Carbonate
CaO	Calcium Oxide
CaOH2	Calcium Hydroxide
DOE	U.S. Department of Energy
EDR	Electrodialysis Reversal
EIA	Energy Information Administration
EOR	Enhanced Oil Recovery

* This is an edited, reformatted and augmented version of National Energy Technology Laboratory, Publication No. DOE/NETL-311/040609, dated November 1 2009.

EPA	Environmental Protection Agency
EPEC	Existing Plants, Emissions and Capture Program
EPRI	Electric Power Research Institute
FGD	Flue Gas Desulfurization
FeS2	Iron Sulfide
GHG	Greenhouse Gas
GIS	Geographic Information System
gph/MW	Gallons Per Hour Megawatt Capacity
gpm	Gallons Per Minute
HERO	High-Efficiency Reverse Osmosis
HPC	Heterotrophic Bacteria
IGCC	Integrated Gasification Combined Cycle
ISGS	Illinois State Geological Survey
LFCM	Ligand Functionalized Core Materials
MAPP	Mid-Continent Area Power Pool
MGD	Million Gallons Per Day
mg/L	Milligrams Per Liter
MW	Megawatt
NERC	North American Electric Reliability Council
NETL	National Energy Technology Laboratory
NH3	Ammonia
NMLRC	National Mine Land Reclamation Center
NPDES	National Pollutant Discharge Elimination System
NaHCO3	Sodium Bicarbonate
O&M	Operations and Maintenance
PAA	Polyacrylic Acid
PBTC	2-Phosphonobutane-1,2,4-Tricarboxylic Acid
PC	Pulverized Coal
pCi/L	Picocuries Per Liter
PMA	Polymaleic Acid
POTWs	Publicly Owned Treatment Works
ppm	Parts Per Million
psi	Pounds Per Square Inch
PSI	Puckorious Scaling Index
R&D	Research and Development
RCRA	Resource Conservation and Recovery Act
RO	Reverse Osmosis
SEM/EDS	Scanning Electron Microscopy/Energy Dispersive Spectrometry
SJGS	San Juan Generating Station
SNL	Sandia National Laboratories
SO2	Sulfur Dioxide
SO4	Sulfate
TDS	Total Dissolved Solids
TKPP	Pyrophosphate
TMDL	Total Maximum Daily Load

TPH	Total Petroleum Hydrocarbons
TSS	Total Suspended Solids
TTA	Tolyltriazole
UND EERC	University of North Dakota Energy and Environmental Research Center
USGS	United States Geological Survey
VOC	Volatile Organic Compounds
WQC	Water Quality Criteria

EXECUTIVE SUMMARY

A U.S. Department of Energy National Energy Technology Laboratory (DOE/NETL) study estimated that in 2005, total U.S. freshwater withdrawals for thermoelectric power generation amounted to approximately 146 billion gallons per day (BGD), while freshwater consumption was 3.7 BGD. Coal-fired power plants use water for cooling water systems, flue gas desulfurization (FGD) makeup, boiler makeup, ash handling, wastewater treatment, and general plant wash down. As the population continues to increase, electric and water demands are expected to grow.

Research examining the energy/water link in coal-based power plants is sponsored by DOE/NETL under the Existing Plants, Emissions and Capture (EPEC) Program. The research and development (R&D) program aims to minimize freshwater withdrawal and consumption without introducing negative impacts of poor water quality on plant operations. In order to develop cost-effective approaches to using lower-quality water sources, the Non-Traditional Sources of Process and Cooling Water component of the EPEC Program includes research focusing on potential new water sources, including brackish and saline water supplies and various domestic and industrial wastewaters. Studies in the treatment of municipal treated water/reclaimed water, produced waters from oil and gas wells, mine pool waters, produced waters from carbon dioxide (CO2) storage in saline formations, and ash pond basins focus on understanding the location, volumes, and quality of different non-traditional waters. This component also develops treatment technologies to reduce the scaling, biofouling, and corrosion potential of available impaired waters that would not otherwise be useful in cooling water systems or other processes within thermoelectric plants.

Reclaimed water (treated municipal wastewater) is widely available in communities throughout the United States in sufficient volumes and is reliable enough to supply power plant cooling water. Reclaimed waters are already being used in more than 50 U.S. power plants and are subject to Federal and state regulations in order to protect worker and public health. Researchers have developed small pilot-scale cooling towers for side-by-side evaluation of the use of impaired waters under different operating conditions. Standard chemical additives provided reasonable scaling and biofouling control. The corrosion rates were not affected by the corrosion inhibitors but were strongly influenced by the degree of scaling on the corrosion coupons. The effects of combinations of tertiary treatments (e.g., nitrification, filtration, carbon adsorption) and different chemical treatment regimens are under investigation.

Produced waters from oil and gas wells, as well as enhanced coalbed methane (CBM) recovery activities, generally exceed discharge limits because chloride and total dissolved solids (TDS) concentrations are very high. With appropriate pretreatment, both a high-efficiency reverse osmosis (HERO) system and a constructed wetland treatment system offer promising treatment techniques to utilize produced waters as process waters. Additional studies are being conducted using nanofiltration techniques and tertiary treatments to utilize produced waters. Ownership of the produced waters is one key regulatory issue that must be addressed, but policy makers also need to investigate methods to allow designations of these waters for beneficial purposes.

Some mine pool waters can provide sustainable waters to power plants but generally have a low pH and high TDS, thus necessitating treatment prior to use in cooling systems. Scaling issues have not yet been ameliorated, and corrosion analyses indicate that mine pool waters should not be used where aluminum is present in the system. However, chloramine and tolyltriazole (TTA) effectively inhibited biofouling and corrosion on steel and copper in the pilot-scale cooling towers. Some studies are also focused on characterizing the water flow rates and quality in particular mine drainages to understand the potential uses as process and cooling waters. In addition, investigations into these mines as direct wide area heat sinks shows lower capital costs but higher operating costs than traditional cooling methods. Studies of the mine pool waters as geothermal heat pumps show that a single coal seam could heat and cool as many as 20,000 homes.

Waters produced when CO2 is stored in saline formations could provide significant volumes of process cooling water if this practice is widely adopted to reduce greenhouse gas (GHG) emissions. At the San Juan Generating Station (SJGS), the HERO system should be able to recover 83 percent of the water. The reject stream from the HERO process would be treated by the brine concentrator (BC) to recover an additional 73 percent of the reject. The total cost was estimated at $5.32 per 1,000 gallons of treated water. Additional studies are considering the transport and treatment of produced waters from additional saline formations. To treat waters like these, a new study will couple electrodialysis reversal (EDR) to lower TDS with a ligand functionalized core material to remove silica.

Regarding non-traditional waters at thermoelectric power plants, current studies focus on recovering ash pond waters. The studies indicate that untreated ash pond waters typically exceed the National Pollutant Discharge Elimination System (NPDES) permit and water quality criteria for arsenic, cadmium, chromium, copper, mercury, selenium, and zinc. Chemical treatments and constructed wetlands treatment systems have been shown to treat the scaling, biofouling, and corrosion potential of these waters. The wetlands system was also measured to decrease the toxicity and pollutant concentrations (arsenic, chromium, mercury, selenium, and zinc) from the ash pond waters. With regard to legislative and regulatory issues, U.S. Representative Nick Rahall introduced a bill (Coal Ash Reclamation and Environmental Safety Act of 2009) that would impose uniform design, engineering, and performance standards on coal ash impoundments. Also, Senator Barbara Boxer has introduced a resolution calling on the U.S. Environmental Protection Agency (EPA) to utilize its existing authority under the Resource Conservation and Recovery Act (RCRA) to regulate coal combustion waste.

In addition to the treatment technologies for non-traditional waters, NETL is also sponsoring work to allow utilities to assess the availability of suitable water supplies in their area. These mapping tools are being constructed as web interfaces and will access National and state-level data sets for both surface freshwaters and non-traditional waters. Groundwater volumes have not been well characterized in many of these data sets, but the researchers are including information about water quantity and quality where available. These projects are being conducted in conjunction with the Ground Water Protection Council, the United States Geological Survey (USGS), state officials, and other NETL contractors.

SECTION 1. BACKGROUND

U.S. population estimates show an increase of 2.8 million individuals from July 1, 2007, through July 1, 2008.[1] The increases since July 2000 average 0.94 percent per year. As the U.S. population continues to increase, demands for both freshwater and electricity are expected to rise. Thermoelectric power generation requires a large quantity of freshwater to support operations. In regions of the country with limited freshwater supplies, thermoelectric power generation may be competing with other uses for dwindling freshwater supplies. Other issues associated with continued supply of freshwater include:

- All regions of the United States are subject to water shortages, particularly during droughts.
- Regional imbalances in available water may require additional energy.
- Non-consumptive water uses must be explored.
- Long-term societal and economic sustainability (e.g., for areas reliant on slow-charging aquifers) may require current and future limitations on water usage.

Water usage may be measured by both water withdrawal and water consumption. A withdrawal rate refers to the net volume of freshwater that is removed from a water supply over a given time. Water consumption is that portion of water withdrawal that is not returned directly in the discharge (e.g., lost through evaporation to the atmosphere). In other words, the user's consumption refers to the difference between the withdrawal rate and the discharge rate.

In 2005, thermoelectric plants ranked slightly ahead of agricultural irrigation in terms of freshwater withdrawal rates (41 percent), according to the United States Geological Survey (USGS).[2] This percentage is presented to indicate the necessity for water withdrawal by thermoelectric plants. However, the water consumption for thermoelectric power plants represented just 3 percent of the total consumption in the 1995 version of the USGS circular.[3]

Since 2002, the Department of Energy's (DOE) National Energy Technology Laboratory (NETL) has sponsored research examining the energy/water link in coal-based power plants under the Existing Plants, Emissions and Capture (EPEC) Program (formerly known as the Innovations for Existing Plants Program). The research and development (R&D) program aims to minimize freshwater withdrawal and consumption without introducing negative impacts of poor water quality on plant operations. The Water-Energy Interface research focuses on four areas[4]:

1. **Advanced Cooling Technologies** – With the goal to improve performance and reduce costs associated with wet cooling, dry cooling, and hybrid cooling

technologies, this component includes projects dealing with condensing technology evaluation, scale prevention, novel filtration methods, testing an evaporative cooler with impaired waters, prevention of zebra mussel fouling, and development of foam with high thermal conductivity for air cooled condensers.

2. **Water Reuse and Recovery** – This component focuses on reuse of cooling water and the associated waste heat and water recovery from coal and flue gas. The projects include coal drying; desalination and additional power generation from waste heat; condensing heat exchangers, membranes, and liquid desiccants in flue gas streams; and the use of wetlands to treat and reuse waters.
3. **Non-Traditional Sources of Process and Cooling Water** – In order to develop cost-effective approaches to using lower-quality, non-traditional water sources, this component includes studies in the treatment of municipal treated water/reclaimed water, produced waters from oil and gas wells, mine pool waters, produced waters from carbon dioxide (CO2) storage in saline formations, and ash pond basins.
4. **Advanced Water Treatment and Detection Technology** – Focusing on the removal of pollutants, the studies in this component include investigations into a passive integrated treatment facility, the creation of a market-based approach for water quality and carbon emissions, use of novel anionic clay sorbents, and pilot-scale constructed wetland treatment systems.

NETL has set quantifiable metrics for the Water-Energy Interface research projects within the EPEC Program. The short-term goal is to prepare technologies for commercial demonstration by 2015 that, when used alone or in combination, can reduce freshwater withdrawal and consumption by 50 percent or greater for thermoelectric power plants equipped with wet recirculating cooling technology, at a levelized cost of less than $4.40 per thousand gallons of freshwater conserved. The 2020 long-term goal is more ambitious and calls for a freshwater withdrawal and consumption reduction of 70 percent or greater, at a cost less than $2.90 per thousand gallons of freshwater conserved.

This report focuses on the research that has been done to date for the Non-Traditional Sources of Process and Cooling Water component. Studies in this research area focus on identifying potential new water sources, including brackish and saline water supplies and various domestic and industrial wastewaters. In particular, these studies focus on understanding the location, volumes, and quality of different non-traditional waters, as well as treatment technologies to reduce the scaling, biofouling, and corrosion potential of available impaired waters that would not otherwise be useful in a cooling water system or other processes within a thermoelectric plant.

SECTION 2. SCOPE AND ORGANIZATION

This report covers ongoing and recent projects that fall under the DOE/NETL list of projects dealing with uses of non-traditional waters for power plant makeup waters. Where appropriate, some other NETL water-related activities in similar technical areas have also been described to offer a broader perspective to the research for certain non-traditional waters, but this report is focused on non-traditional waters and is not intended to summarize all water-related activities at NETL.

Instead of organization by individual projects, each Section of this report discusses all of the projects that deal with a particular category of non-traditional waters. The entire report is organized to discuss the focal points of the research:

- Section 3: Regulatory and Permitting Considerations
- Section 4: Thermoelectric Power Plant Water Quantity and Quality Needs
- Section 5: Municipal Treated Wastewater/Reclaimed Water
- Section 6: Produced Waters from Oil and Gas Wells
- Section 7: Mine Pool Waters
- Section 8: Produced Water from Carbon Dioxide Storage in Saline Aquifers
- Section 9: Recovered Plant Discharges
- Section 10: Locating Sources of Non-Traditional Waters

Section 3 briefly discusses the regulatory and permitting considerations associated with the use of non-traditional waters. Section 4 discusses the various makeup water needs within a thermoelectric plant for the different operations. Section 5 through Section 9 represents the majority of the presented NETL-funded work on non-traditional waters. Based on the reports from the investigators, each of these Sections presents the environmental concerns associated with the untreated water, a water quality assessment, treatment technology descriptions, performance goals (for studies underway), performance results (for completed studies), and other issues. Table 2-1 summarizes the investigations that are covered in this report. Section 10 describes studies that are underway in an effort to identify nearby sources of non-traditional waters through geographic information systems (GIS).

Table 2-1. Investigations Reviewed in this Report

Section	Investigation	Research Group	Start Date
5	Use of Reclaimed Water for Power Plant Cooling (DE-AC02-06CH11357)	Argonne National Laboratory	--
5, 7, 9	Reuse of Treated Internal or External Wastewaters in the Cooling Systems of Coal-based Thermoelectric Power Plants (NT42722)	University of Pittsburgh and Carnegie Mellon University	March 2006
5	Use of Treated Municipal Wastewater as Power Plant Cooling System Makeup Water: Tertiary Treatment versus Expanded Chemical Regimen for Recirculating Water Quality Management (NT06550)	University of Pittsburgh and Carnegie Mellon University	October 2008
6, 8	Use of Produced Water in Recirculating Cooling Systems at Power Generating Facilities (NT41906)	EPRI at the San Juan Generating Station	July 2004
6	Reuse of Produced Water from CO2 Enhanced Oil Recovery, Coal-Bed Methane, and Mine Pool Water by Coal-Based Power Plants (NT0005343)	Illinois State Geological Survey	October 2008
6	Nanofiltration Treatment Options for Thermoelectric Power Plant Water Treatment Demands (FWP-08-014250)	Sandia National Laboratories	September 2008

Table 2-1. (Continued)

Section	Investigation	Research Group	Start Date
6, 7	Identification of Incentive Options to Encourage the Use of Produced Water, Coal Bed Methane Water, and Mine Pool Water	Argonne National Laboratory	--
6, 9	An Innovative System for Treatment of Non-Traditional Waters for Reuse in Power Generation (NT42535)	Clemson University	July 2005
7	Development and Demonstration of a Modeling Framework for Assessing the Efficacy of Using Mine Water for Thermoelectric Power Generation (NT42723)	National Mine Land Reclamation Center (NMLRC) of West Virginia University	March 2006
7	Strategies for Cooling Electric Generating Facilities Using Mine Water: Technical and Economic Feasibility (NT41908)	West Virginia University	September 2003
7	Use of Mine Pool Water for Power Plant Cooling	Argonne National Laboratory	--
8	Study of the Use of Saline Aquifers for Combined Thermoelectric Power Plant Water Needs and Carbon Sequestration at a Regional Scale (07-013812)	Sandia National Laboratories	July 2007
8	Thermoelectric Power Plant Water Demands Using Alternative Water Supplies: Power Demand Options in Regions of Water Stress and Future Carbon Management (FWP-08-014053)	Sandia National Laboratories	June 2008
8	Technology to Facilitate the Use of Impaired Waters in Cooling Towers (NT0005961)	GE Global Research	October 2008
10	Optimization of Cooling Water Resources for Power Generation (FT40320-01.24 and NT43291-05.1)	University of North Dakota Energy and Environmental Research Center (UND EERC)	June 2007 July 2008
10	Internet Based, GIS Catalog of Non-Traditional Sources of Cooling Water for Use at America's Coal-Fired Power Plants (NT0005957)	Arthur Langhus Layne, LLC	October 2008

This document is intended to be regularly updated as additional research is conducted by investigators. The research reports through October 2009 are reported in this version.

3. Regulatory and Permitting Considerations

The purpose of this Section is to highlight the broad issues associated with regulations and permitting that must be considered when proposing to use non-traditional waters within a thermoelectric power plant. Most of the NETL research reports on non-traditional waters cover this topic, at most, in a cursory fashion, but access to a water supply in a water-scarce region is limited by law, regulation, and/or policy.

With regard to regulatory and permitting considerations, the principal issues of concern include plant discharge from treatment processes, laws related to water rights, and regulations governing water reuse. In addition, one researcher lists available government incentives to promote the use of non-traditional waters. These topics are covered in the following sections.

Plant Discharge from Treatment Processes

Federal and state laws apply to the plant discharges, but vary depending on the composition of the non-traditional water source as well as the treatment processes. The Federal Clean Water Act (CWA) establishes authority over plant discharges into the environment, but most state water pollution control agencies are responsible for administering and enforcing the provisions of the CWA.

Under the CWA, plant discharges into navigable waters are regulated by permits issued under the National Pollution Discharge Elimination System (NPDES). The NPDES permits are based on both technology-based effluent limits and water quality-based effluent limits, whichever is more stringent. When discharges are made to designated impaired waters, the effluent limits specified in the permit are more stringent. In addition to pollutant concentration limits, NPDES discharge regulations may also restrict the total maximum daily load (TMDL) of a particular pollutant or the quality of the receiving water body (e.g., minimum dissolved oxygen concentrations).

Regulations may also govern the effluent discharge flow rate from a power plant. In the U.S. Environmental Protection Agency's (EPA) Guidelines for Water Reuse[5], examples for regulating flow in order to preserve existing habitat (e.g., Endangered Species Act) and to meet demand for reclaimed water by communities in the area are cited. The EPA guidelines focused on water reclaimed from municipal wastewater treatment systems and on pollutants associated with wastewater treatment.

In addition, some treatment processes may generate solid or sludge byproducts requiring offsite storage. The Federal Resource Conservation and Recovery Act (RCRA), as well as state laws, will govern requirements associated with these process streams.

Water Rights

Water rights allow the diversion of water and its use for particular purposes. State statutes, regulations, and case law govern the rights of private parties and government entities regarding the use of the public waters.[5] Water rights in Western states (and areas that are water-limited) are generally governed under the appropriative doctrine, and the riparian doctrine typically applies in Eastern states. Both doctrines can affect water reuse projects and apply to surface waters, but groundwater rights are most often administered through an appropriative doctrine.

The appropriative rights system grants the user the rights to a particular beneficial use based on a first-come, first-served basis. Under this system, a new user may not diminish the quantity or quality of the water to any more senior users. Late users may not have rights to the water except in the wet season. Under the appropriative doctrine, state law may promote or constrain reuse projects. Even if the user has first rights to the use of effluent, EPA guidelines suggest four cases that may discourage reuse projects:

- Activities that could reduce discharge (e.g., evaporative cooling) and affect downstream users.
- Changes in the point of discharge or the place of use.

- In times of shortage, a hierarchy of use could prevent reclaimed water from being used for industrial processes.
- Reduced withdrawals from the water supply may jeopardize the availability during a shortage if allocations are based on historic usage.

In the East and water-abundant areas, riparian water rights systems are based on land purchases beside the waterways. All landowners whose property is directly adjacent to a body of water have equal right to make reasonable use of that water. However, riparian users are not entitled to use the water such that stream flow or quality is substantially depleted. Unlike the appropriative doctrine, each landowner is assured some water when available, and the riparian doctrine does not allow for storage of water.

Federal water laws may sometimes impact water rights when usage affects water supply in another state, protected Native American tribal land, or another country. The Federal government may claim jurisdiction in state disputes (e.g., claims to water in the Colorado River during shortages). In addition, Federal reserved water rights refer to the water quantity reserved by the Federal government but do not have to be established at the time of the land's acquisition and are not lost due to non-use or abandonment. The Federal reserved water rights may not cause harm to other water users or take priority over existing appropriations.

Regulations Governing Water Reuse

The University of Pittsburgh is currently conducting a research project for DOE/NETL on the treatment and use of various impaired waters in the cooling system for thermoelectric power plants.[6] One of the project tasks was to assess the relevant regulations and permitting issues associated with the use of impaired waters. Although the Federal government has not established regulations that specifically address reuse of impaired waters, a number of states have developed guidelines or regulations. The state regulations pertain primarily to minimizing cooling tower drift, which could possibly pose a health risk to the public. Table 3-1 highlights state regulations from the University of Pittsburgh study related to regulatory requirements for reuse of impaired water in industrial settings.

Table 3-1. Summary of State Regulatory Requirements on Water Reuse[6]

State	Regulatory Requirement
Arizona	AAC, R18-11, Article 3 provides the water standard of reclaimed water quality. R18-11-308 states that reclaimed water quality requirements for industrial reuse applications are industry-specific and determined on a case-by-case basis.
California	California Code of Regulations Title 22 section 60306 specifies the water quality of recycled water for industrial cooling. In a cooling tower, disinfected tertiary recycled water must be oxidized, coagulated, filtered, and disinfected with both total coliform and turbidity limits. When cooling towers are not involved, the recycled water must be oxidized and disinfected and meet coliform limits. In terms of air regulations, a drift eliminator is required when mist is created in the cooling system.

State	Regulatory Requirement
Florida	Florida Administrative Code 62-610.668 specifies reclaimed water quality requirements for once-through and open cooling towers in terms of $CBOD_5$, total suspended solids, fecal coliform, and pH. The removal percentages for $CBOD_5$, total suspended solids, and fecal coliform are also specified.
Hawaii	Hawaii's Department of Health published guidelines for recycled water. Cooling water that does not emit drift should be oxidized and disinfected while limiting fecal coliform and chlorine residue to specific limits. If drift is emitted, the water should also be filtered and meet turbidity limits. High efficiency drift reducers should be used, and the biocide residual should prevent bacterial concentrations exceeding 10,000/ml.
Maryland	The Pittsburgh researchers reported that a State Senate bill was drafted in 2002 for reclaimed water reuse, but no bill had been passed. The Code of Maryland Regulations does not include rules regarding water reuse.[7]
New Jersey	The 2005 NJDEP technical manual —Reclaimed Water for Beneficial Reuse" mentions type IV reclaimed water as industrial water, but no standard has been established. The guidelines in the technical manual indicate chlorine residue and fecal coliform standards, but the total suspended solids are specified in the NJDEP permit for the facility.
North Carolina	Under NC Administrative Code SubSection 02T Section 15A, reclaimed water should be treated and meet tertiary treatment water quality requirements (BOD_5, TSS, ammonia, fecal coliform, and turbidity).
Oregon	According to Oregon Administrative Rules 340-055, reclaimed water from sewage treatment plants may be used as cooling water under certain conditions. Level II or III waters may be used in non-evaporative cooling systems if they meet criteria for total coliform. Level IV waters may be used in evaporative cooling water systems with stricter limits on total coliform as well as turbidity. The user must —demonstrate that aerosols will not present a hazard to public health."
Pennsylvania	Pennsylvania issued a draft version of "Reuse of Treated Wastewater Guidance Manual" in 2005 and a later draft in January 2009.[8] The expanded draft specifies that re-used industrial process water be Class C or better (measured by BOD, TSS, and fecal coliform). Secondary treatment and disinfection are required (and include specifications for chlorine residual or disinfection design dose if ultraviolet light is used).
Texas	Under Texas Administrative Code 30-210, water types are classified whether the public may come in contact with the water (Type I) or not (Type II). Type II waters used for cooling have limits set for BOD_5 and fecal coliform. Type I waters have stricter limits and a requirement that limits turbidity.
Utah	Utah Administrative Code R317-1-4 classifies water types whether the public may come in contact with the water (Type I) or not (Type II). Re-used cooling water requires secondary treatment and disinfection and must meet limits for TSS, BOD_5, pH, and fecal coliform.
Washington	In 2006 and 2007, the Washington legislature revised Title 90 Section 90.46 to revise the reclaimed water use rule (Section 173-219 WAC) by 2010. Currently an informal comment process is underway on a draft rule.[9] The water quality standards are currently set by the 1997 version of the Water Reclamation and Reuse Standards. If mists are not created, Class C water may be oxidized and disinfected with limits on total coliform, chlorine residual, BOD, TSS, and turbidity. If mists are created, only Class A water that has been oxidized, coagulated, filtered, and disinfected may be used with stricter standards on total coliform than the non-mist case.

Table 3-1. (Continued)

State	Regulatory Requirement
Wyoming	Under Wyoming's Standards for the Reuse of Treated Wastewaters Section 21,[10] Section 12 states that treated wastewater is authorized for reuse for irrigation purposes only. However, Section 7(b) states that —The determination of the manner in which treated wastewater is to be reused is a local determination.‖ Standards are set for fecal coliform.

As indicated in Table 3-1, states' regulatory requirements are focused on the reuse of wastewaters (primarily municipal) but not on other non-traditional waters, as evidenced by the concerns about microorganisms in cooling tower drift. Any additional rules or regulations that researchers have identified regarding their particular types of non-traditional waters are described in the Sections that deal with the individual waters.

Available Governmental Incentives

Gillette and Veil[11] conducted an NETL study on potential incentive options to encourage industrial sources to use produced water, coalbed methane (CBM) water, and mine pool water. In that report, the authors list several reasons why different levels of government might offer incentives (e.g., reducing demand on conventional sources, enticing industry, and eliminating contamination threats). They then identified and briefly described five governmental incentive measures that could be applied:

1. Direct grants to pay full or partial costs for constructing and/or operating a facility.
2. Tax/royalty subsidies or reductions by local or state governments (set to expire after a specific period of time).
3. Reduced water costs to the user.
4. Assured market for a private company that collects, treats, and distributes these water resources to ensure an appropriate return on the investment.
5. Regulatory relief where possible on environmental requirements, water quality restrictions, permitting, etc.

Section 4. Thermoelectric Power Plant Water Quantity and Quality Needs Introduction

The purpose of this Section is to provide an overview of the makeup water quantity and quality requirements for thermoelectric power plants. In the future, thermoelectric power plants may be required to supplement and/or replace freshwater makeup with some source of non-traditional water in order to conserve freshwater resources. However, some sources of non-traditional water could have physical or chemical properties that would be detrimental to power plant operations without appropriate treatment. Therefore, the water quality requirements for power plant systems need to be specified in order to determine appropriate treatment requirements for the non-traditional water.

Background

Thermoelectric power plants – coal, oil, natural gas, and nuclear fueled power generators using a steam turbine based on the Rankine thermodynamic cycle – require significant quantities of water for generating electrical energy. The largest demand for this water is process cooling. The two commonly used metrics to measure water use are withdrawal and consumption. The water required for thermoelectric plant operation is withdrawn primarily from large volume sources, such as lakes, rivers, oceans, and underground aquifers.[12] Water consumption is used to describe the loss of withdrawn water, typically through evaporation into the air, which is not returned to the source. USGS estimated that thermoelectric generation accounted for approximately 41 percent of freshwater withdrawals, ranking slightly ahead of agricultural irrigation as the largest source of freshwater withdrawals in the United States in 2005.[13] However, the corresponding water consumption associated with thermoelectric generation accounted for only 3 percent of total U.S. freshwater consumption in 1995.[14] A recent DOE/NETL study estimated that in 2005, the total U.S. freshwater withdrawals for thermoelectric power generation amounted to approximately 146 billion gallons per day (BGD), while freshwater consumption was 3.7 BGD.[15]

There are two general types of wet cooling system designs used for thermoelectric power plants: once-through and wet recirculating.[16] Plants equipped with once-through cooling water systems have relatively high water withdrawal, but low water consumption. Conversely, compared to once-through systems, plants equipped with wet recirculating systems have relatively low water withdrawal but high water consumption. In once-through systems, the cooling water is withdrawn from a local body of water and the warm cooling water is subsequently discharged back to the same water body after passing through the surface condenser. Due to the large volume of water required for once-through cooling systems, chemical treatment is generally not practical with the possible exception of chlorination to control microbiological fouling. This Section focuses on water treatment requirements for wet recirculating cooling systems. Figure 4- 1 shows a simplified schematic of a coal-fired power plant equipped with a wet recirculating cooling system.

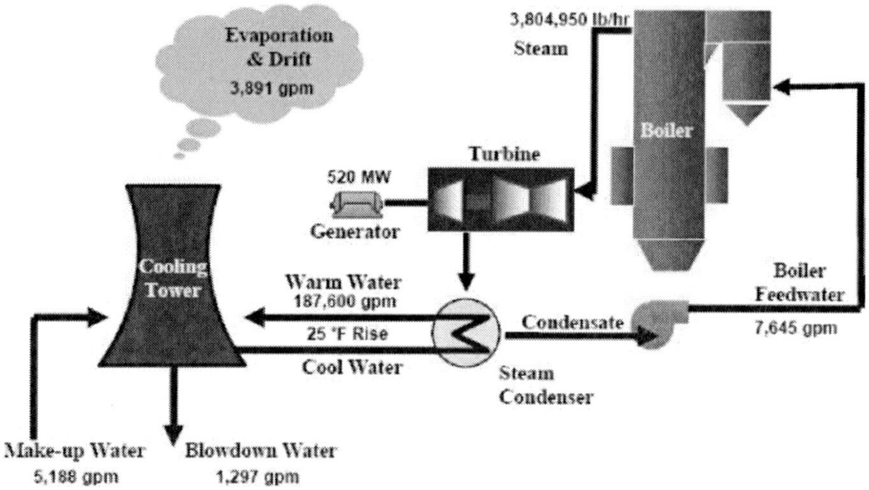

Figure 4-1. Schematic of a Wet Recirculating Cooling Water System.

In wet recirculating systems, the warm cooling water is typically pumped from the condenser to a cooling tower where the heat is dissipated directly to ambient air by evaporation of the water and heating the air. The cooling water is then recycled back to the condenser. Because of evaporative losses, a portion of the cooling water needs to be discharged from the system – known as blowdown – to prevent an excessive buildup of minerals and sediment in the water that could adversely affect performance. The quantity of blowdown required for a particular cooling water system is determined by a parameter known as cycles of concentration (COC), which is defined as the ratio of dissolved solids in the circulating water to that in the makeup water. As the COC increases, the quantity of blowdown and makeup water decreases. In addition to the cooling system, coal-fired power plants use water for flue gas desulfurization (FGD) makeup, boiler makeup, ash handling, wastewater treatment, and general plant wash down.[17] For coal-fired power plants with a wet FGD system, it is the second largest user of makeup water. Table 4-1 shows the estimated makeup water requirements for a 520-megawatt (MW) subcritical coal-fired power plant equipped with a wet recirculating cooling tower operating at four COC and a wet FGD system.[18] Flow rates are given in terms of gallons per minute (gpm) and gallons per hour per megawatt capacity (gph/MW). The cooling system requires a net makeup water flow of 5,150 gpm to replace the evaporation and blowdown water losses less the boiler blowdown water that is directed to the cooling tower. The FGD system requires a net makeup water flow of 571 gpm to replace evaporative losses in the flue gas and FGD gypsum by-product water losses. Water requirements for a supercritical plant would be approximately 10 percent to 12 percent less than those for a subcritical plant on a megawatt basis due to greater plant efficiency.

Table 4-1. Estimated Cooling and FGD System Makeup Water Requirements for a 520-MW Subcritical Coal-Fired Power Plant[18]

Flow Path	Flow Rate, gpm	Flow Rate, gph/MW
Cooling System		
Evaporation	3,891	449
Blowdown	1,297	150
Gross cooling tower makeup	5,188	599
Boiler blowdown to tower	38	4
Net cooling tower makeup	5,150	595
FGD System		
Flue gas evaporation	490	57
FGD gypsum by-product	81	9
Net wet FGD makeup	571	66

As previously mentioned, evaporative losses from the cooling tower can lead to an excessive buildup of total dissolved solids (TDS) in the circulating water system that could adversely affect performance of the steam condenser and cooling tower. Blowdown of a portion of the circulating water is the primary method used to control TDS concentrations. The COC can be calculated using the following formula (assuming zero drift losses):

COC = (E + B)/B

Where:

COC = Cycles of concentration
B = Blowdown rate, gpm
E = Evaporation rate, gpm

The makeup water quality requirements for the cooling water system are contingent on the water quality requirements of the circulating water and the COC that the system is operated. For example, if the maximum acceptable level of chlorides in the circulating water system is 750 milligrams per liter (mg/L) and the system is operated at three COC, then the makeup water would need to be limited to 250 mg/L (750 divided by three).

Conversely, the makeup water quality can dictate the allowable COC for which the cooling system can be operated. The formula above can be re-written to calculate the required blowdown rate (B) based on the COC and evaporation rate (E):

B = E/(COC-1)

For example, if the circulating water's maximum chloride concentration were 1,000 mg/L and the makeup water chloride concentration was 250 mg/L, then the maximum COC would be four (1,000/250). Assuming the evaporation rate was 3,000 gpm, the required blowdown rate would be 1,000 gpm (3,000/4-1), and total makeup water would be 4,000 gpm (3,000+1,000). However, if the makeup water chloride concentration increases to 500 mg/L, then the maximum COC would be two (1,000/500), the required blowdown rate would increase to 3,000 gpm (3,000/2-1), and total makeup water would increase to 6,000 gpm (3,000+3,000). If such an increase in the blowdown rate or makeup were not acceptable, then either the makeup water would require treatment to reduce the concentration of chlorides, or the circulating water would require treatment to accommodate the increased chloride concentration.

Water Quality Impacts on the Cooling Water System

Without proper control, the physical and chemical characteristics of the makeup and circulating water can lead to scale formation, corrosion, or microbiological fouling that adversely affect cooling water system performance.

Scale Formation

As water evaporates from the circulating water system, dissolved solids from the makeup water can accumulate to saturation levels and begin to precipitate out of solution as solid scale-forming deposits. Scale formation is a function of the chemical composition of the makeup water, circulating water temperature and pH, and COC.[19] There are several chemical species (including calcium carbonate [$CaCO_3$], calcium sulfate, calcium phosphate, and magnesium silicate) that contribute to TDS in freshwater makeup to the cooling system that can form scale deposits if allowed to reach saturation levels. Recent studies of various non-traditional waters have identified $CaCO_3$, silica, barium sulfate, and calcium sulfate to be chemical constituents that could potentially limit COC based on the effective treatment limits of commercially available scale control technologies.[20]

Calcium carbonate is the principal scale-forming component in the cooling water system and is formed by the decomposition of calcium bicarbonate. There can also be a problem with calcium sulfate scaling if sulfuric acid is used to control calcium bicarbonate levels. The potential for $CaCO_3$ scaling is dependent on a number of inter-related physical and chemical characteristics of the cooling water system. Although blowdown is the primary method used to control TDS concentration in the circulating water system, some form of treatment may also be required (e.g., chemical additives and/or various methods of filtration). Chemical treatment using phosphonates or polymers are used to prevent $CaCO_3$ scaling by acting as a dispersant and/or crystal modifier.

Corrosion

Corrosion in the cooling water system occurs primarily due to electrolytic action. Therefore, an increase in TDS raises conductivity and the potential for corrosion. Excessive chloride and sulfate ion concentrations are of particular concern regarding corrosion potential. However, the acceptable level of water quality to minimize corrosion is dependent on the materials of construction used throughout the cooling water system. The most serious concern with corrosion occurs with the tubing and tubesheet of the steam condenser. These components are typically constructed of copper alloys, stainless steel, or titanium. The copper alloys are the most susceptible to corrosion, but the stainless steels are also at risk. For example, ammonia (NH_3) can cause corrosion to copper and copper alloys.[21] Pitting corrosion of stainless steels can also be caused by manganese oxide in the circulating water.[19] Chemical treatment using various corrosion inhibitors can be used to prevent corrosion. Copper alloys can be treated with tolyltriazole, and stainless steels can be treated with ortho-phosphate.[22]

Microbiological Fouling

Microbiological growth within the circulating water system results in biofilm deposits of slime and algae on heat transfer surfaces. The biofilm deposits can both restrict heat transfer and promote corrosion. Certain chemical species in the makeup water – such as nitrogen, phosphate, and organic compounds – can promote microbiological growth within the circulating water system.[19] Microbiological fouling can be controlled by chlorine, bromine, sodium hypochlorite, chlorine dioxide, hydrogen peroxide, ozone, or various proprietary chemical treatments.[19,23,24]

Water Quality Requirements for the Cooling Water System

As previously discussed, proper water quality requirements must be maintained in order to prevent scale formation, corrosion, or microbiological fouling that could adversely affect cooling water system performance. Although corrosion and microbiological fouling might be more severe when using non-traditional water for makeup, it is likely that scale formation will be the greatest concern. The maximum COC using the non-traditional water needs to be determined based on a review of calculated saturation ratios of potential mineral deposits compared to the practical treatment limits of commercially available scale control technologies. The following subsections provide recommended water quality requirements for cooling water systems as recommended by various industry experts. Since several methods of treatment may be required, it is important that any interaction between the treatments does not impede their performance. Therefore, a qualified cooling water system consultant should be used to assure proper water treatments are being utilized. Table 4-2, Table 4-3, and Table 4-4 provide cooling water chemistry limit recommendations from three organizations. SPX Cooling Technologies, Inc. is one of the largest U.S. suppliers of wet and dry cooling systems for thermoelectric power plants. Table 4-2 presents SPX general recommendations for circulating water quality requirements.[23] Nalco is one of the largest U.S. vendors for cooling system water treatment technologies used in the power generation industry. Table 4-3 shows the concentrations or saturation ratios for constituents of non-traditional water that Nalco assumes as upper limits when using commercially available scale inhibitors.[20,25] The Electric Power Research Institute (EPRI) conducted a study for the California Energy Commission (CEC) on the use of degraded water sources for cooling water makeup in power plants. Table 4-4 presents a summary of the cooling water criteria developed by EPRI for the CEC study.[26]

Table 4-2. SPX General Cooling Water Chemistry Limits

Constituent	Cooling Water Chemistry Limit
pH	Range of 6 to 8
Chloride	Less than 750 mg/L measured as NaCl
Sulfate (SO_4)	Less than 1,200 mg/L
Sodium bicarbonate ($NaHCO_3$)	Less than 200 mg/L
Calcium carbonate ($CaCO_3$)	Langelier saturation index ≈ 0 Ryznar stability index between 6 and 7

Table 4-3. Nalco Water Chemistry Limits for Cooling Water

Constituent	Cooling Water Chemistry Limit
Iron	Up to 5.0 mg/L
Silica	Up to 200 – 250 mg/L depending on temperature and magnesium concentration
$CaCO_3$	Saturation ratio up to 250
Calcium sulfate	Gypsum saturation ratio up to 4.5
Barium sulfate	Barite saturation ratio up to 900

Table 4-4. EPRI Water Chemistry Limits for Cooling Water

Constituent	Units	Cooling Water Chemistry Limit
Ca	mg/L $CaCO_3$	< 900 (Note 1)
Ca w/ PO_4 present	mg/L $CaCO_3$	20 – 285 (depending on pH and PO_4 concentration) (Note 1)
Ca x SO_4	$(mg/L)^2$	< 500,000 (Note 1)
Mg x SiO_2	mg/L $CaCO_3$ x mg/L SiO_2	< 35,000 w/o scale inhibitor < 75,000 w/ scale inhibitor (Note 1)
Alkalinity (HCO_3 + CO_3)	mg/L $CaCO_3$	30 – 50 w/o scale inhibitor 200 – 250 w/ scale inhibitor (Note 1)
SO_4	mg/L	Case specific limit (Note 1)
SiO_2	mg/L	< 150 (Note 1)
PO_4	mg/L	Case specific limit (Note 1)

Table 4-4. (Continued)

Constituent	Units	Cooling Water Chemistry Limit
Fe	mg/L	< 0.5 (Note 1)
Mn	mg/L	< 0.5
Cu	mg/L	< 0.1
Al	mg/L	< 1
S	mg/L	< 5
NH_3	mg/L	< 2 (Note 2)
pH	mg/L	6.8 – 7.2 w/o scale inhibitor 7.8 – 8.4 w/ scale inhibitor (Note 1)
pH w/ PO_4 present		7.0 – 7.5 (Note 1)
TDS	mg/L	< 70,000 (Note 1)
TSS	mg/L	< 100 with film fill < 300 with splash fill
Langelier index	--	< 0
Ryznar index	--	> 6
Puckorius index	--	> 6

Note 1: Limit represents a conservative value. EPRI recommends software calculation of saturation level. Note 2: <2 mg/L NH_3 limit applies when copper alloys are present in cooling water system. Does not apply to 70-30 or 90-10 copper nickel alloys.

Water Quality Impacts on the FGD System

Proper makeup water quality is also necessary to maintain process performance and minimize corrosion and scaling in FGD systems. The basic FGD process involves spraying an alkaline reagent into the coal combustion flue gas that reacts with sulfur dioxide (SO_2) and water to form a precipitated salt by-product that can be filtered from the system and either disposed in an impoundment/landfill, or marketed for beneficial use. Lime or limestone can be used as the reagent and the salt by-product is calcium sulfite and/or calcium sulfate. The FGD process can be further defined as either dry or wet depending on the amount of water used to spray the reagent into the flue gas. Figure 4-2 is a schematic of a wet limestone FGD absorber module showing the major process flows. The majority of U.S. wet FGD systems use limestone, while most dry FGD systems use lime as the reagent. However, there are a number of U.S. plants that have wet FGD systems that use a magnesium-enhanced lime as the reagent.

This discussion will focus on the water quality requirements for wet lime and limestone FGD systems because they require greater volumetric flow rates of makeup water. There are several processes in wet FGD systems that require makeup water including reagent preparation and mist eliminator wash. Makeup water is also required for pump seals and to maintain process density and/or tank levels. These water uses are necessary for operation of the FGD process and help replenish the water losses associated with evaporation to the flue gas and the chemical formation and disposal of the FGD waste by-product. In addition to using freshwater for FGD makeup, some plants utilize cooling tower blowdown. Reclaimed FGD process water from the FGD solids dewatering process can also be used for some of the FGD-related processes. Water may also need to be purged from the FGD system to control the build-up of chlorides. This serves a similar function to blowdown in the cooling water system.

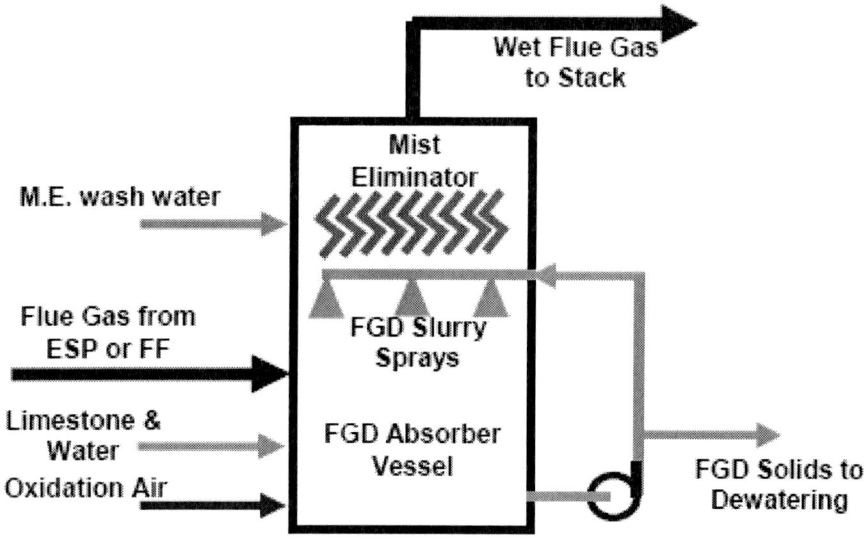

Figure 4-2. Wet Limestone FGD Process Schematic.

In a wet lime FGD system the lime feed slurry is prepared by mixing dry pebble lime with freshwater to convert the calcium oxide (CaO) to calcium hydroxide (Ca[OH]2) – known as lime slaking. The slaked lime is then further diluted with either freshwater or FGD process water to produce lime slurry with the proper solids concentration. Similarly, in a wet limestone FGD system, limestone slurry is typically prepared by grinding dry pebble limestone to a fine powder in a wet ball mill where freshwater or reclaimed process water is added to provide proper dilution. The FGD absorber vessel includes a mist eliminator device to prevent excessive carryover of slurry droplets in the exit flue gas stream. The mist eliminator chevrons are washed with freshwater to prevent the accumulation of solids that could lead to pluggage.

The relative amount of calcium sulfite/sulfate formed in a wet FGD system is a function of several process parameters, including the amount of oxygen available in the flue gas to convert the sulfite to sulfate. A majority of today's wet FGD systems use forced oxidation to further drive the FGD chemistry to maximize calcium sulfate production. The use of forced oxidation to promote calcium sulfate formation has two primary FGD process benefits: 1) there is less scaling of internal FGD components, which increases performance and reliability; and 2) dewatering and disposal of calcium sulfate is easier and less expensive than calcium sulfite. However, perhaps just as important, calcium sulfate – also known as gypsum – can be used as a low-cost alternative to naturally-occurring rock gypsum in the production of wallboard. As a result, many coal-fired power plants equipped with forced oxidation wet FGD systems have been selling their FGD synthetic gypsum by-product to wallboard manufacturers, which both eliminates the cost of disposal and provides additional revenue.

Water Quality Requirements for the FGD System

Lime and limestone wet FGD systems have different water quality requirements for both reagent preparation and mist eliminator wash due to differences in process chemistry. According to Carmeuse, a major U.S. supplier of FGD lime and limestone reagents, while lime preparation requires water with relatively low sulfate content (less than 500-1,000 mg/L), limestone preparation should be relatively unaffected by high TDS water quality. Conversely, wet lime FGD systems are likely to be unaffected by high TDS mist eliminator wash water, but wet limestone FGD mist eliminators could be adversely affected. However, for either lime or limestone systems, mist eliminator wash water should be low enough in calcium and sulfate to maintain gypsum saturation levels below 30 percent to minimize the potential for scaling.[27]

Although wet FGD systems are designed with appropriate materials of construction to address a highly corrosive environment, the relatively high chloride content of some non-traditional water could adversely impact some stainless steels.[28] For example, EPRI established a 5,000-mg/L limit for chloride levels at the San Juan Power Station to minimize corrosion of stainless steel components of the wet FGD absorber.[29]

In 1996, Radian International (now part of URS Group) prepared an FGD process design manual for DOE that included recommendations on water quality requirements.[30] The following information is taken from the manual:

1. In addition to using freshwater, much of the required wet FGD makeup water can be met using relatively low-quality water sources such as treated wastewater, ash sluice water, cooling tower blowdown, or treated municipal wastewater.
2. Trace species, such as iron and manganese, can act as oxidation catalysts. This would be a benefit to forced-oxidation systems, but a detriment to inhibited-oxidation systems. However, iron and manganese may be a concern if the gypsum is being marketed for wallboard production.
3. For wet lime FGD systems, the lime slaking water must be relatively low in TDS to minimize unwanted chemical reactions that could reduce slaking efficiency. Therefore, freshwater is preferred for this application.
4. For wet limestone FGD systems, the limestone grinding operation does not require a chemical reaction; therefore, reclaimed FGD process water can be used for limestone slurry preparation without affecting reagent reactivity.
5. Mist eliminator wash water for magnesium-lime and inhibited-oxidation limestone wet FGD systems can be reclaimed FGD process water, provided the calcium sulfate relative saturation[31] is less than 50 percent and the water is free of suspended solids. However, it is recommended that freshwater be used for mist eliminator wash in forced-oxidation limestone wet FGD systems.
6. Pump seal water should be freshwater.
7. High levels of chloride can concentrate in the FGD process water depending on the chlorine content of the coal and the FGD system water balance. Excessive levels of chloride can decrease SO2 removal performance in limestone FGD systems and accelerate corrosion of stainless steel components. Chloride concentrations should be kept below 15,000 mg/L to prevent any significant impact on SO2 removal. The chloride concentration should be kept below 3,000 mg/L for Type 316L stainless

steel, less than 6,000 to 8,000 mg/L for Type 317L stainless steel, and less than 10,000 mg/L for Type 904L stainless steel.

8. Wallboard manufacturers have quality control specifications for the physical and chemical parameters of FGD synthetic gypsum. It will be important to assure that any use of impaired water for FGD makeup not adversely impact the quality of the synthetic gypsum.
9. Depending on the overall water balance of the wet FGD system and the need to limit the buildup of chlorides and TDS in the process liquor, it may be necessary to discharge water from the system similar to a cooling tower blowdown. The FGD discharge water may need treatment in order to meet the plant's NPDES permit requirements.

SECTION 5. TREATED MUNICIPAL WASTEWATERS/RECLAIMED WATERS

This Section covers Existing Plants Program-funded work on the use of treated municipal wastewaters (also called reclaimed waters) for power plant cooling. Reclaimed waters represent a large alternative source of water that is being used in a variety of ways throughout the United States. Many studies have been published on the use of reclaimed water for irrigation, fire protection, wildlife habitat enhancement, and industrial applications. EPA has issued several reports on reclaimed water and water reuse, the most comprehensive of which was completed in 2004. EPA (2004)[32] documents numerous examples of reclaimed water reuse, including some involving cooling water at power plants. A national organization, the WateReuse Association, sponsors conferences and meetings that provide information on many aspects of the reuse of reclaimed water. While there has been a significant amount of work devoted to the study of the use of reclaimed water, a small percentage of that has been devoted to its use in utility applications.[33]

There are several reasons why reclaimed water represents a potentially valuable alternative source of cooling water. Municipal wastewater is available in communities throughout the country, and treatment facilities are designed to handle specific design flows. Thus, the quantity of water leaving the treatment plants is well defined and can be factored into the design and operation of power plant cooling systems. Data analysis has revealed that 81 percent of power plants proposed for construction by the Energy Information Administration (EIA) would have sufficient cooling water supply from one to two publicly owned treatment works (POTWs) within a 10-mile radius, while 97 percent of the proposed power plants would be able to meet their cooling water needs with one to two POTWs within 25 miles of these plants. Thus, municipal wastewater will be the impaired water source most likely to be locally available in sufficient and reliable quantities for power plants.[34]

This Section reflects work that has been, or is currently being done, in the following NETL studies:

- "Use of Reclaimed Water for Power Plant Cooling," (DE-AC02-06CH11357) conducted by Argonne National Laboratory (ANL) to develop information on the actual use of reclaimed water for cooling or other purposes at power plants.[33]

- "Reuse of Treated Internal or External Wastewaters in the Cooling Systems of Coal-based Thermoelectric Power Plants," (NT42722) conducted by the University of Pittsburgh and Carnegie Mellon University to assess the potential of using three different impaired waters (secondary treated municipal wastewater, passively treated coal mine drainage, and ash pond effluent) as cooling water in coal-based thermoelectric power generation.[34]
- "Use of Treated Municipal Wastewater as Power Plant Cooling System Makeup Water: Tertiary Treatment versus Expanded Chemical Regimen for Recirculating Water Quality Management," (NT06550) conducted by Carnegie Mellon University and University of Pittsburgh to compare the impacts of tertiary treatment and chemical treatments on the performance of treated municipal wastewater in power plant cooling systems.[35]

Water Quality

Reclaimed water is treated municipal wastewater, and as such is subject to several sets of state and Federal regulatory requirements. EPA (2004)[33] contains a detailed discussion of the national and state requirements associated with the full spectrum of reuse of reclaimed water. A brief summary is provided below.

The CWA requires that all discharges of pollutants to surface waters (streams, rivers, lakes, bays, and oceans) must be authorized by a permit issued under the NPDES program. EPA has the authority to administer the NPDES program, but it can also authorize states to implement all or parts of the program. Municipal wastewater treatment facilities must obtain an NPDES permit before discharging treated wastewater.

The national minimum technology-based discharge standard for municipal wastewater is referred to as "secondary treatment." The secondary treatment regulation places limits on five-day biochemical oxygen demand (BOD5), total suspended solids (TSS), and pH. The pH limit is that the effluent must be maintained between six and nine. For BOD5 and TSS, the 30-day average values cannot exceed 30 mg/L, the seven-day average values cannot exceed 45 mg/L, and the 30- day average percent removal cannot be less than 85 percent.

In addition, NPDES permits can include water quality-based limits that consider state water quality standards and the available dilution of the receiving water body. These limits may be set on metals, organic compounds, nutrients, whole effluent toxicity, or other parameters. Bacteria and other microbial pathogens that could potentially be released to the atmosphere through a cooling tower plume are of particular concern. Most NPDES permits for municipal wastewater discharges include limits for microbial pathogens and/or controls for disinfection. However, these limits are not based on any Federal regulations, but rather on state regulations and policies.

Most states have either regulations or guidelines regarding the reuse of treated municipal wastewaters. Regulations refer to actual rules that have been enacted and are enforceable by government agencies. Guidelines, on the other hand, are generally not enforceable but can be used in the development of a reuse program or for exercise of agency discretion.

State requirements typically consider the probable degree of public exposure to reclaimed water. In circumstances where public exposure is unlimited, reclaimed water must be highly treated. In addition to the general and secondary standards, states may require more stringent biochemical oxygen demand (BOD) and TSS standards. To control microbial pathogens, states usually establish limits on fecal or total coliform bacteria, and may require that the wastewater be filtered before it can be reused as reclaimed water. States also frequently impose turbidity standards to ensure that the treatment systems are working properly. Chlorine is the most commonly used disinfectant. Some plants use ultraviolet radiation, ozone, or other halogenated compounds. If the disinfectant creates corrosive conditions (e.g., chlorine in some applications), operators may dechlorinate prior to use.

Table 5-1 summarizes some of the limits that are commonly applied by states to the use of reclaimed water.[33]

Table 5-1. Commonly Applied Limits for Water Quality Constituents in Reclaimed Water

Water Quality Constituent	Limit
Fecal Coliform	<200/100 ml
Five-Day Biochemical Oxygen Demand (BOD_5)	<30 mg/L
pH	6.0 – 9.0
Residual Chlorine	>1 mg/L
Total Suspended Solids (TSS)	<30 mg/L

EPA (2004)[32] includes suggested guidelines for reusing treated municipal wastewater in various applications. The elements of the guidelines related to using reclaimed water for power plant cooling include the water quality constituent concentrations cited in Table 5-1, along with the following:

- The minimum treatment requirements include secondary treatment and disinfection with, if needed, chemical coagulation and filtration. Additional treatment may be performed to prevent scaling, corrosion, biological growth, fouling, and foaming.
- Windblown spray should not reach areas accessible to workers or the public. This requirement can be met by providing a setback distance of 90 meters. This setback may be reduced or eliminated if a high level of disinfection is provided.

One other requirement relates to marking and identification of reclaimed water systems. The general standard is to paint any pipes, valves, pumps, and other components that convey reclaimed water a bright purple color to distinguish them from other types of water. Figure 5-1 shows reclaimed water pipes and pumps painted purple at a power plant in Maryland.[33]

Current Use of Reclaimed Waters for Power Plant Cooling

Argonne (2007)[33] developed a database of power production facilities using reclaimed water for cooling based upon information found in Federal and state databases along with internet searches and detailed investigative work. The full database of 57 facilities is attached as Appendix A. Table 5-2 through Table 5-5 provides summary information from that database.

Table 5-2 shows the distribution of facilities by state. Florida, California, Texas, and Arizona have the highest number of facilities. All of these states have experienced freshwater shortages for many years. Their use of reclaimed water is a direct response to water shortages. However, reclaimed water is being used throughout the country. Several facilities are found in states that traditionally have been thought to have ample water supplies (Massachusetts, Maryland, Mississippi, New Hampshire, New Jersey, Pennsylvania, and Wisconsin).[33]

Figure 5-1. Reclaimed Water Pipes and Pumps Painted Purple.

Table 5-2. Geographic Distribution of Facilities[33]

State in Which Power Plant and Wastewater Treatment Plant are Located	Number of Power Plants Using Reclaimed Water
Arizona	3
California	13
Colorado	1
Florida	17
Iowa	1
Kansas	1
Maryland	2
Massachusetts	3
Mississippi	1
Nevada	2
New Hampshire	1
New Jersey	2
New Mexico	1
Pennsylvania	1
Texas	7
Wisconsin	1

Table 5-3 shows the volumes of reclaimed water used by power production facilities. The data are well distributed and range from less than 0.1 million gallons per day (MGD) to as high as 55 MGD. Many facilities use seasonal-varying rates, as the power plants shift loads to meet power demand.

Table 5-3. Volume Usage Distribution[33]

Volume of Reclaimed Water Used (MGD)	Number of Facilities
<0.1	6
0.1– 0.5	10
0.51– 1.0	12
1.01– 5.0	15
5.01– 10.0	6
>10.0	5

Table 5-4 shows the varying uses of reclaimed water at the power plants. Most of the plants use the reclaimed water as cooling tower makeup. Other uses include cooling pond supplementation, air scrubbing, boiler feed water, and source water to maintain underground pressure in geothermal fields in Northern California.[33]

Table 5-4. Distribution of Facilities by Their Use of Reclaimed Water[33]

Use of Reclaimed Water	Number of Facilities
Air scrubbers	2
Boiler feed water	1
Cooling ponds	2
Cooling tower makeup	46
Cooling tower makeup and scrubber water	1
Injected to increase pressure at geothermal fields	2

Table 5-5 shows the time period in which power facilities began using reclaimed water. Most power plants in the database began using reclaimed water after 1990. The oldest reported usage came from two power plants in Southern California that started using reclaimed water in 1967.[33]

Table 5-5. Distribution of Facilities by Time Period in Which Power Plant Began Using Reclaimed Water[33]

Time Period in Which Power Plant Began Using Reclaimed Water	Number of Facilities
1960– 1969	2
1970– 1979	3
1980– 1989	8
1990– 1999	18
After 2000	19

Treatment of Reclaimed Water

Some of the chemical constituents commonly found in reclaimed water may lead to operational issues. For example:

- Calcium phosphate and other constituents can cause mineral scaling.
- Corrosion, pitting, and stress cracking damage to metal heat transfer surfaces and structural metal surfaces can result from the presence of a variety of constituents (e.g., NH3 can damage copper, copper alloys, and other metals).
- BOD, phosphate, and NH3 can lead to biofouling of heat transfer surfaces and excessive biological growth on cooling tower fill material surfaces.

There are a variety of ways for power plant operators to control the water quality and cycles of concentration. Some of the concentrated water can be removed from the system as cooling tower blowdown. In order to replace the water lost to evaporation, blowdown, and drift, new makeup water is added to the system. Operators can adjust the flow volumes of blowdown and makeup to maintain suitable chemistry in the recirculating water. Another method for controlling water quality and increasing the cycles of concentration is to treat the incoming reclaimed water. Many different treatment processes can be used to target those constituents that are likely to limit the cycles of concentration or that could damage components of the cooling system, as described below.[33]

Reclaimed water is treated at the municipal wastewater treatment plant to at least secondary treatment standards in order to comply with Federal requirements. In some states (e.g., California), reclaimed water must be treated to tertiary treatment levels prior to reuse. However, other states do not place additional requirements on the quality of reclaimed water. In most cases, if the reclaimed water is not already treated beyond secondary standards, the power plant using the water will either treat it at the power plant site or pay the wastewater treatment utility to provide additional treatment.[33]

Table 5-6 shows the starting treatment levels and the additional treatment provided for reclaimed water used for power plant cooling at eight facilities. The starting level of treatment ranges from secondary to tertiary. The additional treatment steps include chemical addition, clarification, disinfection, pH adjustment, and biological treatment.[33]

Other Issues

The overall objective of the University of Pittsburgh/Carnegie Mellon University study[34] is to assess the potential of using three different impaired waters (secondary treated municipal wastewater, passively treated coal mine drainage, and ash pond effluent) as cooling water in coal-based thermoelectric power generation. The study is assessing geographic proximity, pretreatment requirements, available quantities, and regulatory and permitting issues that are relevant for application of these impaired waters in a cooling system. In addition, key design and operating parameters that would ensure successful use of impaired waters without detrimental impact on the performance of the cooling system are being evaluated. Another important project objective is the development and demonstration of small pilot-scale cooling towers for side-byside evaluation of the use of impaired waters under different operating conditions. The pilot-scale cooling towers have been built, deployed, and operated at the Franklin Township Municipal Sanitary Association located in Southwestern Pennsylvania, as pictured in Figure 5-2.

The pilot-scale towers were operated for two consecutive 25-day periods. During the first run, secondary treated effluent was fed to the pilot-scale cooling towers. During the second run, two of the towers were run with secondary treated effluent for additional testing while the third tower was run with a tertiary effluent (secondary effluent subjected to nitrification and granular media filtration) feed stream. Chemical treatment agents were added to the incoming waters to evaluate their impacts on cooling tower performance. These included scaling inhibitors polymaleic acid (PMA) and 2-phosphonobutane-1,2,4-tricarboxylic acid (PBTC), corrosion inhibitors tolyltriazole (TTA) and pyrophosphate (TKPP), and chlorine compounds (free chlorine and monochloramine) to combat biofouling. The target concentrations of chemical treatment agents used in the analyses are presented in Table 5-7.

Detailed information on tower operations during the runs was recorded. Solids deposition rates on stainless steel coupon surfaces, corrosion weight loss of metal alloys, and heterotrophic planktonic/sessile bacteria were documented during both runs. Numerous water chemistry parameters were monitored to obtain detailed understanding of the cooling tower behavior. It was documented that the towers were able to perform according to design specifications and adequately simulate the operation of full-scale cooling towers.

Table 5-6. Examples of Treatment Processes for Power Plant Cooling Water at Eight Specific Facilities[a,32]

Power Facility and Location	Average Cooling Water Supply and Return Flow	Treatment Processes at Wastewater Treatment Plant	Treatment of Cooling Water at Power Plant
Lancaster County Resource Recovery Center, Marietta, PA	Supply = 0.65 MGD Return = 0 Zero discharge; all blowdown is evaporated or leaves plant as sludge.	Secondary treatment with alum, flocculation and polymer addition followed by settling and phosphorus removal	Clarification and addition of ferric sulfate, polymer, and sodium hypochlorite
PSE&G, Ridgefield, NJ	Supply = 0.3 to 0.6	Secondary treatment	Water chemistry is
	MGD. The blowdown is disposed of with plant wastewater to local sewer system.		controlled with biocide, neutralization, and surfactant
Hillsborough County Solid Waste to Energy Recovery Facility (operated by Ogden Martin Corp.) Tampa, FL	Supply – 0.7 MGD (includes some use for irrigation water). The blowdown of 0.093 MGD is mixed with plant wastewater and returned to the WWTP.	Advanced treatment with high-level disinfection. Partial tertiary treatment removes phosphorus.	Addition of chlorine, biocide, surfactant, trisodium, phosphate, and sulfuric acid
Nevada Power – Clark and Sunrise Stations, Las Vegas, NV	Supply = 2.72 MGD average to Clark Station Return = 0 Blowdown is discharged to holding ponds for evaporation.	Advanced secondary treatment with nitrification, dentrification, and biological phosphorus removal; tertiary treatment using dual-media filtration and disinfection with chlorine	None at present time; previously had been treated with lime and softener
Panda Brandywine, Brandywine, MD	Supply = 0.65 MGD Blowdown is returned to the WWTP.	Secondary treatment, biological nutrient removal, sand filters	Addition of corrosion inhibitors, sodium hypochlorite, acid, and antifoaming agents

Table 5-6. (Continued)

Power Facility and Location	Average Cooling Water Supply and Return Flow	Treatment Processes at Wastewater Treatment Plant	Treatment of Cooling Water at Power Plant
Chevron Refineries, El Segundo and Richmond, CA	Supply = 3 to 5 MGD Return = 0	Tertiary treatment at both. El Segundo uses an ammonia stripping process; Richmond uses a caustic soda treatment process	Richmond plant uses a commercial chemical additive [chemistry no specified in EPAQ (2004)]
Curtis Stanton Energy Center, Orange County, FL	Supply = 10 MGD Return = 0 Blowdown is evaporated in concentrator and crystallizer units at the power plant.	Advanced wastewater treatment including filtration, disinfection, and biological nutrient removal.	Addition of acid, scale inhibitor, chlorine; control of calcium
Palo Verde Nuclear Plant, Phoenix, AX	Total supply to three units = 72 MGD Return = 0 All blowdown is evaporated in ponds.	Secondary treatment	Tertiary treatment plant consisting of trickling filters for ammonia removal, clarification fro removal of phosphorus, magnesium, and silica; addition of dispersants, defoaming agents, and sodium.

[a] The information included in this table is reproduced or paraphrased from an earlier study. The information may not match the more current data shown in the database in Appendix A.

Table 5-7. Target Concentrations for Chemical Treatment Agents

Chemical Treatment Agent	Tower A Run 1	Tower B Run 1	Tower C Run 1	Tower A Run 2	Tower B Run 2
TTA	2	1	2	2	0
TKPP	10	0	10	0	0
PMA	10	0	20	10	0
PBTC	5	0	10	0	0
Free Chlorine	1~2	1~2	1~2	--	--
Monochloramine	--	--	--	2~3	2~3

In the analysis of impacts on scaling behavior, for the first run, both PMA and PBTC were added as scaling inhibitors. However, no PBTC was detected in the recirculating water, indicating that PBTC is not effectively maintained in the solution and cannot assist to prevent scaling. Scanning electron microscopy/energy dispersive spectrometry (SEM/EDS) analysis of the scales indicates the presence of both carbonate and phosphate-based solids. In addition, microorganisms were present in the solids due to inadequate biofouling control program in the first run. Only PMA was used in the second run (at the same concentrations), and reasonable scaling control was achieved because the biofouling control was also much more effective.

As indicated in Table 5-7, during the first run, biofouling was controlled by free chlorine levels targeted at 1 to 2 mg/L. The microbial activity in the cooling towers was above the desired level of 10^4 CFU/ml for planktonic heterotrophic bacteria (HPC) whenever the concentrations of chlorine plus monochloramine fell below 1 mg/L. Results from the second run show better control over microbial activity. In the second run, biofouling was controlled by monochloramine levels targeted at 2 to 3 mg/L, so excursions above the 10^4 CFU/ml value for HPC were much less common.

Figure 5-2. Pilot Scale Cooling Towers in Operation at the Franklin Township Municipal Sanitary Authority.[34]

Overall, mild steel, copper, and copper-nickel showed acceptable corrosion rates in the pilot scale towers with or without corrosion inhibitor addition, as indicated in Table 5-8. The corrosion inhibitors TTA and TKPP did not appear to reduce corrosion rates or increase inhibition effectiveness, and scaling seemed to be primarily responsible for protection of the metal alloys. Higher monochloramine levels to prevent biofouling caused higher corrosion rates initially, but their influence was minimized after the scale layer grew thicker and protected the metal alloys. So, when treated municipal wastewater is used in cooling tower systems, it appears that corrosion effects should be considered only after scaling and biofouling controls are determined.

Table 5-8. Average Corrosion Rates (MPY) of Metal Alloys in Pilot Scale Cooling Towers

Metal Alloy	Tower A Run 1	Tower B Run 1	Tower C Run 1	Tower A Run 2	Tower B Run 2
Mild Steel (21-day avg.)	3.35 (fair)	3.51 (fair)	3.28 (fair)	4.78 (fair)	8.21 (poor)
Mild Steel (last 5 days avg.)	1.99 (good)	0.37 (excellent)	3.24 (good)	2.45 (good)	3.51 (fair)
Aluminum (21-day avg.)	2.3, pitting (unacceptable)	1.74, pitting (unacceptable)	1.18, pitting (unacceptable)	6.86, pitting (unacceptable)	13.7, pitting (unacceptable)
Copper (21-day avg.)	ND (excellent)	ND (excellent)	0.03 (excellent)	0.11 (good)	0.08 (excellent)
Copper-nickel (21-day avg.)	ND (excellent)	0.03 (excellent)	ND (excellent)	0.05 (excellent)	0.04 (excellent)

ND: non-detectable.

MPY: mils/year = thousandths of an inch/year.

Corrosion Criteria.

Mild steel piping: 0-1 (MPY) excellent; 1-3 good; 3-5 fair; 5-10 poor; >10 unacceptable Copper alloys: 0-0.1 excellent; 0.1-0.2 good; 0.2-0.3 fair; 0.3-0.5 poor; >0.5 unacceptable Pitting is not acceptable for all alloys.

The recently initiated Carnegie Mellon University/University of Pittsburgh study[35] builds on and extends the previous University of Pittsburgh/Carnegie Mellon University study[34] described above. The overall objective of the new study is to evaluate the benefits and costs of implementing tertiary treatment of the municipal wastewater prior to use in recirculating cooling systems versus an expanded chemical regimen for managing the quality of the cooling water when secondary municipal wastewater is used as makeup. Specific objectives of this research are as follows:

- Determine the benefits and costs of subjecting secondary treated municipal wastewater to additional treatment (to remove NH3 by nitrification, particulates by filtration and BOD by granular activated carbon adsorption or microfiltration) prior to use as cooling water makeup.
- Determine different chemical treatment regimens required to manage cooling water quality for different levels of tertiary treatment ranging from no tertiary treatment to inclusion of nitrification, organic compound removal by adsorption, and filtration.
- Perform comparative life cycle cost analyses for different levels of tertiary treatment versus the corresponding chemical treatment regimens.
- Determine critical economic, technical, and social factors that need to be considered in comparative evaluation of tertiary treatment of makeup water versus cooling water chemistry control by chemical treatment (e.g., future changes in freshwater

availability and cost; value of treated wastewater in the future; social support for water reuse efforts).

CONCLUSION

Reclaimed water (treated municipal wastewater) represents a valuable water resource that can be used for cooling and other purposes in electric power plants. It is widely available in communities throughout the United States in sufficient volumes and reliable enough to warrant consideration as power plant cooling water.

Because reclaimed water is treated municipal wastewater, it is subject to considerable regulation prior to use in a power plant setting. Based upon Federal regulations, all municipal wastewater must be treated to secondary standards before it can be released from a wastewater treatment facility. Some states require further treatment prior to reuse of the reclaimed water. The concern prompting further treatment is to minimize the exposure of workers and the public to microbial pathogens. Typically the regulated parameters include fecal coliform, BOD_5, pH, residual chlorine, and TSS.

Beyond regulatory requirements, most reclaimed waters require some treatment prior to use as cooling waters in order to preclude operational issues within the power plant. Untreated reclaimed water could cause problems within the power plant related to scaling, corrosion, and biofouling. Usually this treatment is done onsite at the power plant, or the utility will pay the municipal treatment plant to perform supplemental treatment. These supplemental treatments can include chemical additions, coagulation, filtration, and clarification among others.

Reclaimed waters are already being used in more than 50 U.S. power plants. Most of the power plants currently using reclaimed water are in states with freshwater shortages such as Florida, California, Texas, and Arizona. However, 12 other states currently have facilities that use reclaimed water for cooling, including states that are typically thought to have abundant freshwater supplies. As competition increases for limited water resources in the future, reclaimed water may supply cooling water to additional electrical utilities.

Ongoing research is being conducted to evaluate the expanded use of reclaimed water in power plant settings. The University of Pittsburgh/Carnegie Mellon University study currently underway is considering factors such as geographic proximity, pretreatment requirements, available quantities, and regulatory and permitting issues that are relevant for application of these impaired waters in a cooling system. In addition, key design and operating parameters that would ensure successful use of reclaimed waters without detrimental impact on the performance of the cooling system are being evaluated. Another important project objective is the development and demonstration of small pilot-scale cooling towers for side-by-side evaluation of the use of impaired waters under different operating conditions. Initial testing of the performance of the pilot-scale cooling towers using municipal wastewater has produced promising results and offers the potential for testing a wide range of treatments to prevent scaling, corrosion, and biofouling.

SECTION 6. PRODUCED WATERS FROM OIL AND GAS WELLS

This Section covers Existing Plants Program-funded work on water produced from oil and gas wells. Although the flow rates for produced waters vary over the lifetime of a well, the water produced from oil and gas extraction activities is significant. In 2001, oil and gas wells in California produced 2.16 billion barrels of water.[36] A low value from onshore oil wells across the Nation was estimated as 14 billion barrels of water per year, with an average water-to-oil ratio of 9.5. The EPRI work[37] cites that seven states generated 90 percent of the produced water in the continental United States in 2002 and that 37 percent of the produced water sources in a USGS database have been deemed treatable (based on TDS less than 30,000 mg/L).

When CBM wells are drilled to reduce hydrostatic pressure on coal, groundwater is also withdrawn. In 2001, the average Powder River Basin well produced 7,400 gallons of water each day and 100,000 cubic feet of methane gas.[36] Projecting 30,000 production wells across Wyoming, the produced water from these sources might be 250 million gallons per day. However, some detailed analyses estimate that water production from most CBM wells will drop rapidly and end in seven to 10 years.

This Section reflects work that has been, or is currently being done, in several NETL studies, including:

- "Use of Produced Water in Recirculating Cooling Systems at Power Generating Facilities," (NT41906) led by EPRI at the San Juan Generating Station (SJGS) that examined treatment options for oil and gas well-produced waters.
- "Reuse of Produced Water from CO_2 Enhanced Oil Recovery, Coal-Bed Methane, and Mine Pool Water by Coal-Based Power Plants," (NT0005343) conducted by the Illinois State Geological Survey (ISGS) that will characterize waters, determine treatments, and develop cost estimates within the Illinois Basin.
- "Nanofiltration Treatment Options for Thermoelectric Power Plant Water Treatment Demands," (FWP-08-014250) by Sandia National Laboratories (SNL) that will examine nanofiltration as an alternative to reverse osmosis (RO) for the treatment of nontraditional waters.
- "Identification of Incentive Options to Encourage the Use of Produced Water, Coal Bed Methane Water, and Mine Pool Water," by ANL that presents scenarios and possible government actions to encourage use of produced waters.
- "An Innovative System for the Efficient and Effective Thermoelectric Treatment of Non-Traditional Waters for Reuse in Power Generation," (NT42535) by Clemson University that tested how wetlands can treat produced waters.

In addition, Section 10 describes additional studies that deal with identifying these sources of water:

- "Optimization of Cooling Water Resources for Power Generation," (FT40320-01.24 and NT43291-05.1) conducted by the UND EERC, which is developing a GIS-based web application decision support system for North Dakota, South Dakota,

Minnesota, and the North American Electric Reliability Council's (NERC) Mid-Continent Area Power Pool (MAPP) region.

- "Internet-Based, GIS Catalog of Non-Traditional Sources of Cooling Water for Use at America's Coal-Fired Power Plants," (NT0005957) conducted by Arthur Langus Layne, LLC that is building national geospatial databases related to non-traditional water sources within 10 miles of power plants and internet-based applications to access the information.

Enhanced oil recovery (EOR) and enhanced CBM recovery activities inject CO2 to extract additional fossil fuels and are generally conducted because of financial incentives for fuel recovery. In addition to fuel extraction, this activity sequesters CO2 in geologic formations, thereby reducing emissions to the atmosphere and helping to mitigate climate change. Section 8 considers the produced water resulting from geologic sequestration of CO2 in the Nation's saline formations which offer substantially more CO2 storage capacity than that achieved via EOR.

Environmental Concerns for Untreated Waters

According to the Clemson study, the composition of the waters extracted from oil and gas wells often does not meet discharge standards. That study compared published literature concentrations for produced waters with the NPDES discharge limits (Table 6-1). The pH values for the produced waters ranged from 7.84 to 9.56.

Table 6-1. Composition of Produced Waters and NPDES Discharge Limits[38]

Constituent	Concentration Range (mg/L)	Minimum/Maximum Discharge Limits (mg/L)
Chemical Oxygen Demand	11.2 – 45.8	170 – 292
Solids, Total Dissolved (TDS)	668 – 38,938	500
Solids, Total Suspended	24 – 101	30 – 70
Total Carbon	3.9 – 1077	5
Total Phosphorus	0 – 0.763	1
Total Sulfate	<10 – 1100	
Polyaromatic hydrocarbons	0 – 0.11	2.8 – 12.2
Naphthalenes	0 – 1.18	0.2
Phenols	0.15 – 21.5	
Benzene	0.683 – 12.1	0.06 – 2.88
Toluene	1.01 – 19.8	0.028 – 5.0
C_2-Benzene	0.05 – 3.7	
Oil and grease	0 – 78	15
Aluminum	0.0015 – 0.473	
Arsenic	0 – 51	
Barium	0.23 – 17	154
Beryllium	0 – 0.001	

Table 6-1. (Continued)

Constituent	Concentration Range (mg/L)	Minimum/Maximum Discharge Limits (mg/L)
Boron	0.019 – 28	
Bromine	0.7 – 350	
Cadmium	0 – 0.098	
Calcium	1 – 818	
Chloride	133 – 108,968	5000
Chromium	0 – 0.39	0.1 – 100
Cobalt	0.0009 – 0.058	
Copper	0 – 1.45	0.37 – 1.0
Gallium	0 – 0.003	
Gold	0 – 0.095	
Iodine	0 – 3.91	
Iron	1.6 – 768	0.6 – 9.4
Lithium	0 – 64	
Magnesium	0.6 – 3671	
Manganese	0 – 15	1 – 4.4
Mercury	0 – 0.086	0.004
Molybdenum	0 – 0.003	
Niobium	0 – 0.002	
Nickel	0 – 1.7	2
Potassium	0.67 – 669	
Rubidium	0 – 1.59	
Ruthenium	0 – 0.018	
Selenium	0 – 0.14	
Silicon	0.05 – 2.9	
Silver	0.012 – 0.15	
Sodium	7 – 13,704	
Strontium	0.65 – 917	
Sulfur	0 – 1556	
Tellurium	0 – 0.004	
Tin	0 – 0.001	50
Titanium	0 – 4402	
Uranium	0 – 0.018	
Vanadium	0 – 1.1	
Zinc	0.017 – 1.6	0.4 – 5.0

Table 6-1 shows that some larger concentrations (e.g., chlorides and TDS) exceed the discharge limits by more than an order of magnitude. Many of these concentrations also exceed toxicity thresholds set under the EPA water quality criteria and cannot be discharged without treatment.

Access and Collection

Investigators considered using existing pipelines and rights-of-way to transport the produced waters to SJGS.[37] The abandoned Hart Canyon Line (previously an oil pipeline) and the abandoned CO_2 Gas Line (previously used in a technique to displace methane from coal) were both identified as possible lines for future use, and an effective delivery system was designed (Figure 6-1) after considering several options. The Hart Canyon and CO_2 Gas Lines would accept both direct injection into the pipelines and delivery by trucks at satellite collection stations. Discussions with the largest producer in the area suggested that the most practical scheme would be to tie these two lines together with a new extension line to the Collection Center.

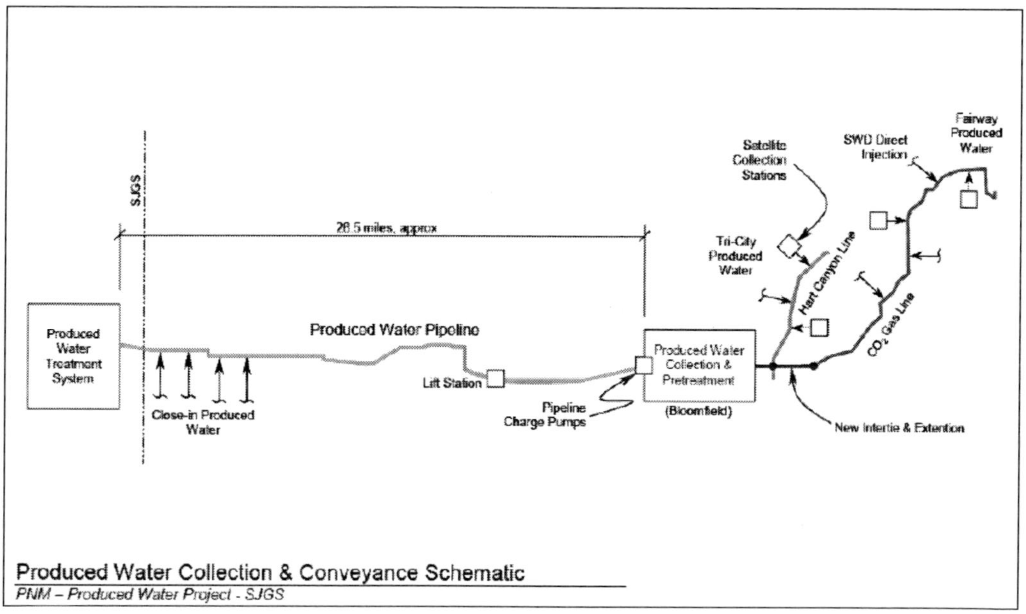

Figure 6-1. Collection Network Diagram Chosen for SJGS Produced Waters.

The Collection Center in Bloomfield would conduct pretreatment to remove oil and grit and equalize chemistry; it would also monitor water quality prior to the transport to SJGS in order to maintain the pipeline integrity. The 14-inch produced water pipeline is composed of high-density polyethylene and would run 28.5 miles using the Public Service of New Mexico transmission right-of-way. A single lift station would be required, as well as charging pumps and clean-out stations along the length. Close-in produced waters (from CBM production) only require simple filtration before addition to the pipeline.

Non-Traditional Water Quality Assessment

Table 6-1 presented the constituents found in produced waters, and some of that information (e.g., TDS concentration ranges) shows that most produced waters are not suitable for direct use as non-traditional water sources. For example, the TDS measured at the McGrath salt water disposal facility varied from 6,400 to 22,600 mg/L in the EPRI study,[37] changing with the locations within the basin from which the water was withdrawn. Chloride concentrations ranged from 2,771 to 12,507 mg/L over the 30-day sampling program at McGrath salt water disposal facility. At the nearby SJGS, the water quality target for TDS in the ash system was only 2,000 mg/L, and the targets for the chlorides in the cooling tower and absorber systems were 1,000 and 5,000 mg/L.

TDS, chlorides, and silica are important concentrations to consider for the non-traditional waters that will be used in plant operations. However, additional water characteristics must be considered when designing a water treatment facility. For example, oil and grease must be removed from the system to prevent fouling of RO membranes. To prevent mineral scale fouling of the membranes, metals (e.g., calcium and magnesium) and silica levels must be reduced.

Prevention of fouling by particulate matter, colloidal matter, and biological films are additional concerns based on the composition of the produced waters near SJGS. The study found that SO2 absorbers represented the least costly use for treated produced water at SJGS.

Produced waters from oil and gas wells (as well as CBM) usually have low concentrations of volatile organic compounds (VOC).[37] Common chemicals in the VOC category (benzene, toluene, ethyl benzene, and xylene) are collectively referred to as BTEX.

Gillette and Veil[36] report that radium may be a concern because many samples have average concentrations exceeding 50 picocuries per liter (pCi/L). Some sources report radium concentrations as high as 3,000 pCi/L, but U.S. nuclear facilities are limited to discharges of 60 pCi/L.

The ISGS project will investigate the produced water quantity and quality available in the Illinois Basin.[39] The samples will be tested for pH, ionic strength, turbidity, TSS, TDS, total petroleum hydrocarbons, total organic carbon, total inorganic carbon, total nitrogen, metal ions, silica, boron, and selected anions (e.g., chloride, sulfate, carbonate, and bicarbonate).

Treatment Technology Descriptions

Four NETL studies have, or are, investigating treatment technologies and their effectiveness on produced waters. The EPRI work with RO systems, Sandia's study with nanofiltration, and the ISGS project all aim to treat waters for use in power plants. The constructed wetlands study by Clemson University compared its results against discharge criteria.

The EPRI work investigated the use of both conventional RO and high-efficiency RO (HERO) in combination with brine concentrators (BCs).[37] Investigators evaluated five treatment alternatives (conventional RO, HERO, BCs only, conventional RO with a BC, and HERO with the same BC). The HERO process with a BC acting on a blend of produced and purge waters was determined to be the most economically feasible approach. The HERO process pre-treats the water to remove scale-forming minerals (e.g., calcium) and other ions associated with high salinity (e.g., chloride). The subsequent BC separates the reject water from the HERO process into mineral-rich and mineral-lean streams. The recovered mineral-lean water from the BC and the HERO permeate would be fed to the cooling towers.

Like conventional RO techniques, the HERO process uses standard spiral-wound polyamide membranes. It differs from conventional techniques because it operates at high pH values (9.5- 10.5) after pre-softening to remove hardness. At pH levels above 10, soluble silica (H4SiO4) dissociates and does not form scale on the membranes.

In addition, pretreatments will be necessary in order to ensure proper operation of RO systems. These include precipitation softening or scale inhibitors to prevent mineral scale from fouling membranes, filtration to remove particulate matter, periodic membrane cleaning with non-oxidizing biocides to remove biofilms, a combination of gravity separation, air flotation, and filtration to remove oil, and neutralization of charged colloidal matter with polymers. The high pH of the HERO system dissolves organic fouling and certain oil constituents to minimize fouling by these mechanisms. An illustration of the pretreatments and the HERO system is shown in Figure 6-2.[37] Similarly, the conventional RO system involved pretreatment through softening (to reduce calcium, magnesium, barium, strontium, and dissolved iron), ultrafiltration (to remove particulate matter and organic foulants), basket strainers (to protect the ultrafiltration device from particulate loading), sodium hypochlorite addition as a disinfectant to protect membranes, and a degasifier to remove CO2 after the pH is adjusted.

Following the exit from the HERO system, the RO reject stream would be sent to the BC for further water recovery under the most economical treatment alternative. The BC would operate in the unseeded mode at high pH (10-11), eliminating the need for acid, antiscalant, and calcium chloride.

RO processes represent a mature technology with a high salt rejection rate (greater than 95 percent). However, nanofiltration membranes require lower pressures to generate equivalent water fluxes at the expense of a lower salt rejection rate (roughly 85 percent). Nanofiltration has typically been used for water softening or the rejection of pathogens and high molecular weight compounds. However, SNL will investigate if TDS in a typical produced water can be reduced from 14,500 mg/L to 2,050 mg/L at a feed pressure of 256 pounds per square inch (psi).[40] With the same feed, model calculations suggest that RO membranes would require 460 psi to lower the TDS from 14,500 to 580 mg/L.

The nanofiltration investigation aims to operate on waters with modest TDS (less than 5,000 mg/L) with very high recovery (roughly 90 percent) and will estimate the associated costs of operation. Nanofiltration techniques are more tolerant of fouling than RO and have a high reject rate for divalent cations. The pilot study will be conducted using Dow Filmtec membranes at a ConocoPhillips CBM production site in the San Juan Basin of northwest New Mexico.

The Illinois Basin study will investigate combinations of conventional treatment techniques, as well as innovative water treatment concepts for treating EOR-produced waters.[41] The conventional techniques include: gravimetric and de-oiling unit operations; filter media including sand, biomass, organoclay, and coal; precipitation and coagulation by adding lime, iron sulfide, and lime followed by iron sulfide; adsorption with granular activated carbon, ion exchange resins, and ion exchange clays; and membrane separation. The de-oiling and filter media processes will not be tested for those waters without oil products (e.g., CBM waters).

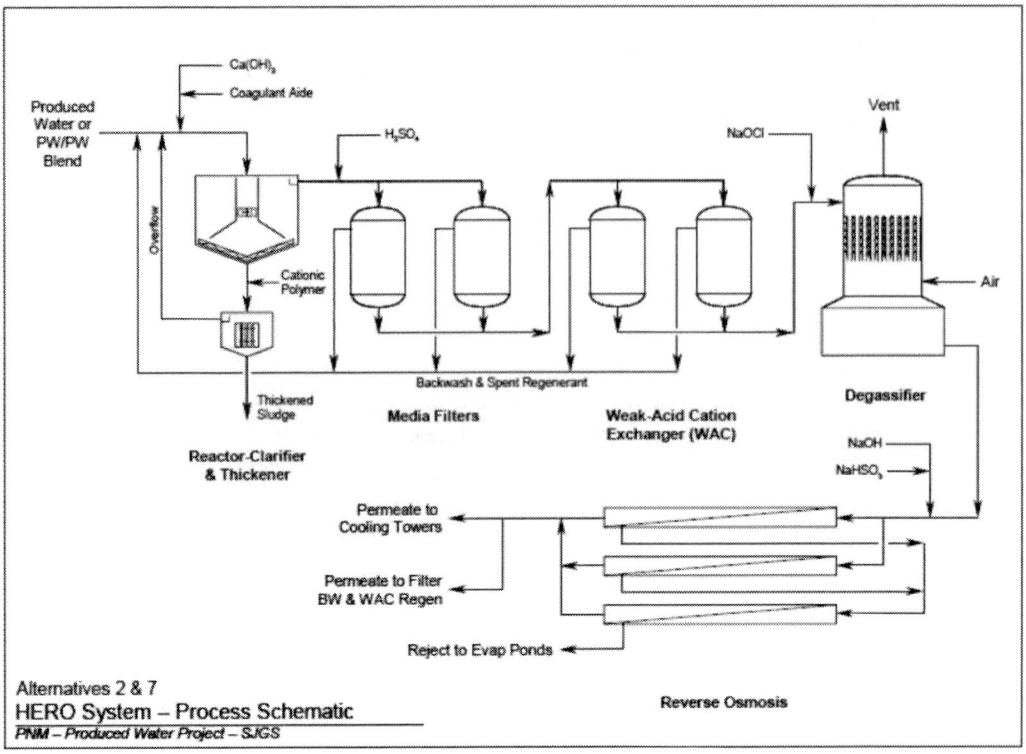

Figure 6-2. Illustration of Pretreatments and HERO System for Produced Waters.[37]

The Illinois Basin study will also consider innovative water treatment technologies being developed at the university, including: system-level integration of desalination, organic separation, and decontamination; catalytic destruction of oxyanions; and adsorbents and photocatalysts with high capacity to adsorb and transform organic and toxic compounds.

The constructed wetland treatment systems (CWTS) designed at Clemson University were tailored to meet the unique needs of the produced waters. In order to design effective sustainable systems, the produced waters were first divided into four categories based on the chloride concentrations:

- Fresh (under 400 to 2,500 mg/L)
- Brackish (2,500 to 15,000 mg/L)
- Saline (15,000 to 40,000 mg/L)
- Hyper Saline (over 40,000 mg/L)

Before introduction to the CWTS, the waters with chloride concentrations greater than 4,000 mg/L were pretreated in a saltwater wetlands system followed by passage through an RO system (Figure 6-3). The RO system was used to lower chloride and TDS concentrations so that levels would be tolerable for freshwater wetland and aquatic life.

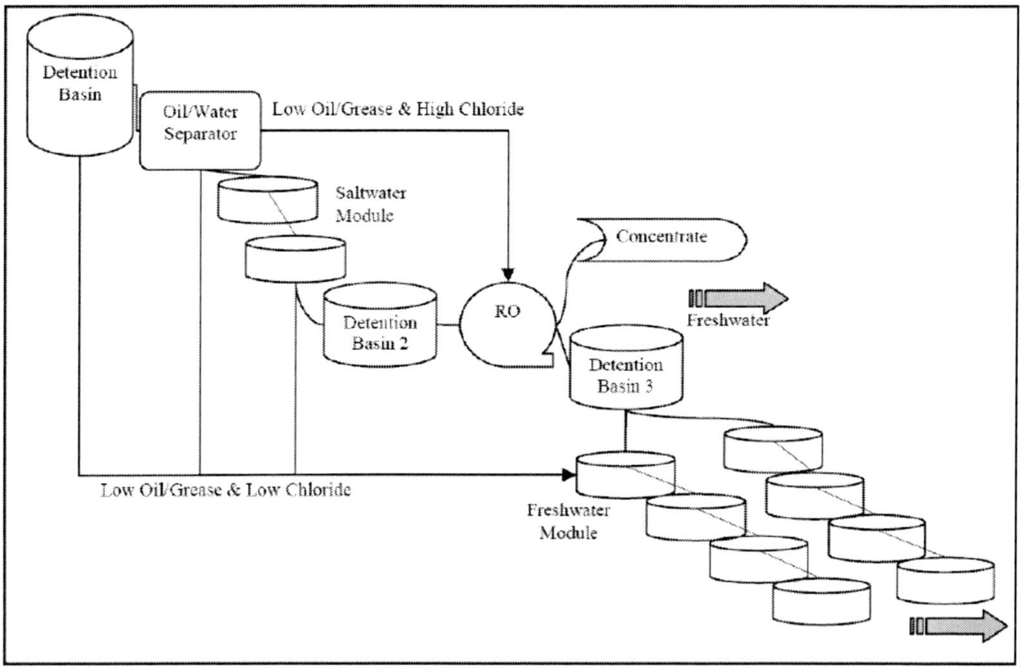

Figure 6-3. Illustration of Constructed Wetlands Treatment System for Produced Waters.[38]

Water passed through four freshwater wetlands reactors, with a 24-hour residence time in each. The first two freshwater wetlands systems contained reducing hydrosoil (river sand hydrosoil and pine mulch) and California bulrush. The final two reactors contained oxidizing hydrosoil (quartz sand) and broadleaf cattail. Concentrations of six targeted pollutants (chlorides, cadmium, copper, lead, zinc, and oil/grease) were tracked through each reactor using simulated waters in each of the four saline categories.

The EPRI study also examined an emerging technology that could significantly reduce the level of oil and particulate matter through ceramic membrane filtration. A traditional difficulty with the use of ceramic membranes for produced waters had been sustaining flux while maintaining separation efficiency. Under previous work, CeraMem Corporation developed a proprietary ceramic membrane and cleaning technique to maintain process flux for several hundred hours between cleanings. Treating pre-filtered produced water with ceramic membranes might eliminate the need for ultrafiltration or RO cartridge filtration.

Performance Goals

Two NETL studies are beginning to investigate the effectiveness of treatment technologies for produced waters. Sandia's study with nanofiltration and the ISGS project to treat waters from the Illinois Basin have each set performance goals to treat the waters so that they can be used at power plants.

The Illinois Basin study will consider options to treat EOR-produced waters at four levels for: (1) coal slurry preparation in an integrated gasification combined cycle (IGCC), (2) FGD makeup water in pulverized coal (PC) plants, (3) cooling tower makeup water in PC and

IGCC, and (4) boiler feedwater in PC and IGCC. The appropriate levels will be determined as one of the early tasks in the investigation before treatment technologies are chosen.

The Illinois Basin treatment technologies will focus on separating sand/grit and oil; removing extremely small oil droplets and precipitates; removing dissolved organic and inorganic compounds through precipitation, adsorption, and membrane separation; and destroying oxyanions. Therefore, the performance metrics should focus on reductions in these measurements.

The nanofiltration study is conducting a pilot test at the ConocoPhillips CBM production site followed by pilot testing at an existing cooling tower system located at SNL. Sandia will test and gather operational data for the nanofiltration technology at the ConocoPhillips site to compare against the model calculations (reducing TDS from 14,500 to 2,050 mg/L at a feed pressure of 256 psi). Instead of investigating treatment of produced waters, the upcoming pilot work at SNL will focus on nanofiltration as a method to reduce the concentration of dissolved solids in the recirculating stream for the cooling tower.

The goals of the wetland treatment study and the EPRI study are discussed in the section below along with their performance results.

Performance Results

The constructed wetland treatment study employed four simulated produced waters (fresh, brackish, saline, and hyper-saline) and measured toxicity, chlorides, cadmium, copper, lead, zinc, and oil/grease. In addition, the study used the Puckorius Scaling Index (PSI) to rate the potential for the waters to form chemical scale and/or cause corrosion. The PSI values of the four simulated produced waters were all in the "Intolerable Corrosion" range, but after CWTS treatment the ratings were reduced to "Heavy Corrosion."

All of the inflow waters were toxic using *C. dubia* in seven-day static renewal toxicity tests. The output waters from the treated fresh, saline, and hyper-saline waters resulted in no mortality or reproductive effects, but the output from the brackish water treatment impaired reproductive processes. The investigators speculated that the poor performance for the brackish water might be attributed to the lack of maturity of the CWTS and relatively high initial loading of constituents of concern.

Most of the cadmium, lead, and zinc were removed in the first two reactors. Although the saltwater wetland module was designed to remove residual organics, it also effectively removed metals from the stream. Overall, the removal rates ranged from 38 to 99.6 percent for cadmium, 91 to 99.8 percent for copper, 93 to 99.3 percent for lead, and 40 to 99.8 percent for zinc. The removal performance was lowest for the brackish waters. The inlet streams contained 1.5 mg/L of total petroleum hydrocarbons (TPH) as diesel fuel (measure of oil/grease), but no TPH was detected in the outlet streams beyond the oil/water separator. The investigators concluded that produced waters could be reused as service water if the pretreatment technologies sufficiently reduced the chloride concentrations.

The EPRI study evaluated the costs of different alternatives and concluded that the treatment of a produced water/purge water blend with the HERO+BC system would have a net cost of $4.52 per thousand gallons, considerably lower than the other systems considered for SJGS. The study also showed that pretreatment was critical to prevent membrane fouling.

The HERO+BC system had the lowest capital cost ($14.1 million) and operating cost ($2.98 million per year) but would require the highest power (1,915 kW).

In evaluation of the CeraMem membrane filtration as a pretreatment, six runs were conducted with Teflon and silica membrane materials. The addition of surfactant dramatically improved the flux stability while providing low TSS in the permeate. The ceramic membrane system was estimated to cost $3.9 million installed and have a unit operating cost of $0.051 per barrel ($1.62 per kgal).

In terms of plant usage, the HERO permeate could all be used in the SJGS SO2 absorbers, but the purge water rate would have to increase slightly to compensate for the increased chloride levels. To reduce costs further, HERO permeate could be fed to both the absorbers and the ash system.

The EPRI study also examined cost estimates for plant examples in other parts of the Nation. A Southwest coal-fired power plant located 7.5 miles from conventional oil production had installed costs of $37.2 million and operating costs ranging from $0.128 to $0.426 per barrel ($4.06 to $13.52 per kgal). A coal-fired power plant in a Rocky Mountain state located 2.5 miles from CBM production would have total installed costs of $15 million and operating costs ranging from $0.169 to $0.371 per barrel ($5.37 to $11.78 per kgal).

Other Issues

Gillette and Veil[36] examined the non-technical issues associated with the use of produced waters. They noted that ownership of produced waters is one critical issue that water policy and law have not yet addressed. It is not clear whether ownership falls to the oil, coal, or gas producer or to the landowner, state, or another entity.

A New Mexico law was signed in March 2004 that allowed SJGS to treat and utilize produced water as process water within the SJGS boundaries.[37] The law was written in a manner that the New Mexico Oil Conservation Division would consider the water as being disposed of at the electric generating facilities, thereby exempting it from other environmental regulations administered by the Office of the State Engineer.

Gillette and Veil listed possible incentives for using produced waters such as direct grants, tax/royalty subsidies or reductions, reduced water costs to the user, assured markets, and regulatory relief. They then constructed 12 scenarios for produced waters and presented examples of such incentives. Those scenarios would be examined in a further study to assess both quantitative and qualitative metrics for potential incentives.

Federal actions include a recent bill that has passed the U.S. House of Representatives. The Produced Water Act of 2009 appropriates 20 million dollars for the research, development, and demonstration of technologies for environmentally sustainable utilization of produced waters.

Conclusion

Produced waters from oil and gas wells, as well as enhanced CBM recovery activities, represent a significant volume of available water that may be transported to nearby power plants. The produced waters generally exceed discharge limits because chloride and TDS concentrations are high.

Chloride and TDS levels generally prevent untreated use of the waters in any power plant operations as a non-traditional water source. However, several studies have, and are, investigating treatment options and possible uses for these waters. The HERO system (with necessary pretreatments) has been coupled with BCs to achieve a cost of $4.52 per thousand gallons, considerably lower than conventional RO processes. The analysis showed that this stream could be fed to both the SO_2 absorbers and the ash system.

Clemson University investigated the use of CWTS to treat produced waters falling into four different salinity classes. Constructed saltwater wetland systems were necessary with RO membrane pretreatment for the more saline produced waters. Four freshwater reactors with reducing and oxidizing components followed the pretreatment. All of the input streams exceeded toxicity criteria, but the outflow streams generally did not. The oil and grease, cadmium, copper, lead, and zinc removal rates were high, and the investigators concluded that produced waters could be reused as service water if the pretreatment technologies sufficiently reduced the chloride concentrations.

An additional study is being done with nanofiltration techniques as a low-pressure alternative to RO. Investigators are also considering potential treatment technologies for produced waters in the Illinois Basin.

Some regulatory issues also need to be addressed if produced waters are to be considered for use at power plants. Ownership of the water is one key issue, but policy makers also need to investigate methods to allow designations of these waters for beneficial purposes.

SECTION 7. MINE POOL WATERS

This Section covers Existing Plants Program-funded work on the use of mine pool waters. The term *mine pool* has typically been used to characterize a volume of ground water that has accumulated in an underground mine after operations in the mine have ceased. This volume, in some cases, represents the volume of a single mine, but in many cases, it may represent the combined volume of several mines, or a mine complex.[42] Most of the research efforts to date have focused on mine pools in Pennsylvania and West Virginia, but the concepts and principles are applicable to mining areas and operations across the country.

There have been two general areas of focus for the use of mine pool waters. One has been the use of mine pool waters as an alternative source for coal-fired power plant cooling processes. Mine pool waters offer the potential to improve the efficiency of the cooling process due to their cool, constant water temperatures (~50°F). The other area of focus has been the potential use of the underground mines themselves as wide area heat sinks.

This Section reflects work that has been, or is currently being done, in several NETL studies, including:

- "Reuse of Treated Internal or External Wastewaters in the Cooling Systems of Coal-based Thermoelectric Power Plants," (NT42722) conducted by the University of Pittsburgh and Carnegie Mellon University to assess the potential of using three different impaired waters (secondary treated municipal wastewater, passively treated coal mine drainage, and ash pond effluent) as cooling water in coal-based thermoelectric power generation.

- "Development and Demonstration of a Modeling Framework for Assessing the Efficacy of Using Mine Water for Thermoelectric Power Generation," (NT42723) conducted by the National Mine Land Reclamation Center (NMLRC) of West Virginia University that will evaluate the quantity and quality of water associated with mine water drainage to meet cooling requirements at a proposed power plant.
- "Strategies for Cooling Electric Generating Facilities Using Mine Water: Technical and Economic Feasibility," (NT41908) by West Virginia University that identified cost saving alternatives to current coal-fired power plant cooling processes.
- "Identification of Incentive Options to Encourage the Use of Produced Water, Coal Bed Methane Water, and Mine Pool Water," by ANL that presents scenarios and possible government actions to encourage use of produced waters.
- "Use of Mine Pool Water for Power Plant Cooling" by ANL that provides preliminary information about the opportunity to use mine pool waters for cooling and process waters in electric generating facilities.

In addition, Section 10 describes an additional study that deals with identifying these sources of water:

- "Internet-Based, GIS Catalog of Non-Traditional Sources of Cooling Water for Use at America's Coal-Fired Power Plants," (NT0005957) conducted by Arthur Langus Layne, LLC that is building national geospatial databases related to non-traditional water sources within 10 miles of power plants and internet-based applications to access the information.

Resource Characterization

As noted above, the research on the use of mine pool waters has focused on resources in Pennsylvania and West Virginia. Thus, the discussion here will focus on these two states, although states with underground mining operations throughout the Appalachians, the Illinois basin, and even in the West (e.g., Utah) can benefit from this research.

There are two geographically distinct coal mining regions in Pennsylvania. Anthracite coal was mined in the north-central/northeastern part of the state, and bituminous coal is mined in the western portion of the state. Coalbeds lie beneath approximately 15,000 square miles in Pennsylvania. In West Virginia, bituminous coal is mined from approximately 17,000 square miles of land in the northern, western, and southern parts of the state. Figure 7-1 shows the locations of coal resources in both states.[38]

Comprehensive inventories of mine pools or mine pool complexes do not exist for either Pennsylvania or West Virginia. Some data has been generated on specific coal fields in the anthracite region of Pennsylvania. In addition, some evaluations of individual mines in the bituminous regions of each state have been conducted. In spite of the lack of comprehensive inventories, sufficient data do exist to allow for estimates of the potential for using mine pool water from specific mines/areas as cooling water at power plants.[42]

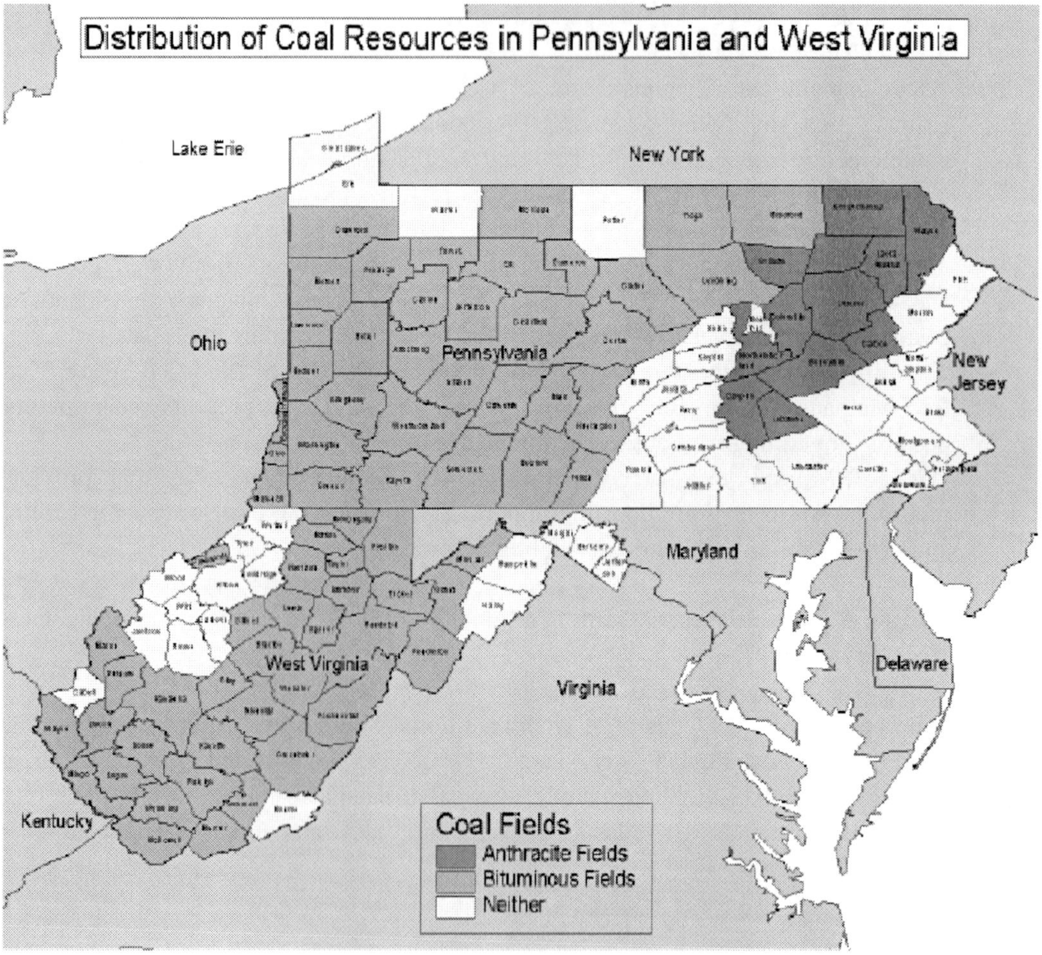

Figure 7-1. Distribution of Coal Resources in Pennsylvania and West Virginia.[42]

For example, a study mapping the mines and mine pools of the Pittsburgh coal seam in western Pennsylvania and northern West Virginia has estimated that the combined storage volume of the 130 mines mapped to date is 250 billion gallons. Another study of selected mines in the Western Middle Anthracite Coal Field estimated the water volume for these mine pools at over 47 billion gallons. In yet another study of mines in the Pittsburgh coal seam, water volumes associated with just eight mines were estimated at nearly 55 billion gallons.[43] These volumes could serve as significant water resources.

Having established that the total volume of mine pool waters is substantial, the next question becomes whether the water yield from the mine pools could be sustained at an adequate rate and over a long enough time frame to serve as a reliable cooling water source for a power plant. If water is withdrawn at a rate significantly higher than the rate of ground water recharge to a mine pool, even with a large initial volume, the ability of a mine pool to supply cooling water over a long time period might be jeopardized. One indicator of the potential rate of ground water recharge to a mine pool is the rate of discharge of a mine outfall. Cravotta et al (1999)[44] measured several outfalls at both anthracite and bituminous sites in Pennsylvania. Table 7-1 lists measured discharge volumes from several of the largest outfalls. As indicated, the discharge volumes range from 450 to 35,000 gpm. Closed cycle cooling requires 4,000 to 6,000 gpm for a 400-MW plant; thus, six of the listed discharges could sustainably supply all of the cooling water needs for a plant of this capacity.

Table 7-1. Selected Mine Discharges from the Anthracite and Bituminous Coal Fields of Pennsylvania[42]

Coal Type	Site	Discharge Volume (ft^3/s)	Discharge Volume (gpm)
Anthracite	Truesdale #2	5	2,240
Anthracite	Bell Water Level Tunnel	22	9,800
Anthracite	Big Mountain Mine	1	450
Anthracite	Derringer Tunnel	13	5,800
Anthracite	Gravity Slope	11	4,900
Anthracite	Jeddo Mine Tunnel	78	35,000
Anthracite	Jermyn Outfall	34	15,000
Anthracite	Old Forge Borehole	78	35,000
Bituminous	S Wilkes Barre Boreholes	32	14,000
Bituminous	Gillespie Run (05)	5	2,240
Bituminous	N. Branch Plum Creek (06)	2	900
Bituminous	Berwind Mine (018)	6	2,700
Bituminous	Little Conemaugh (019)	4	1,800
Bituminous	Browns Run (025)	7	3,100
Bituminous	Sewickley Creek (084)	6	2,700
Bituminous	Tinkers Run (091)	10	4,500

Water Quality

Cravotta et al. (1999)[40] observed that mine pool water in Pennsylvania and West Virginia exhibits a bimodal frequency distribution of pH with two peaks of pH in the range of 2.5 to 4 (acidic) and 6 to 7 (near neutral), as indicated in Figure 7-2. In addition, mine pool water will also have elevated levels of TDS, iron, aluminum, sulfate, and other dissolved metal ions.

Mine pool water becomes acidic from the reaction of oxygen and water with iron-sulfide-bearing minerals found in the coal, such as pyrite (FeS_2). Near-neutral pH mine pool water results from the buffering of mine pool water with calcareous minerals, such as calcite ($CaCO_3$). There are many factors that affect the chemistry (pH, acidity, metals concentration) of mine pool water, including:

- Mineralogy of the coals and overburden.
- Quantity of water flowing through the mine.
- Residence time, path length, and depth of water circulation through the mine. Availability of oxygen in mine water.
- Mine design (e.g., up-dip versus down-dip).
- Active pumping, either within the mine or within the influence of adjacent mines.
- Exposed surface of sulfide minerals.
- Presence or absence of calcite minerals.[42]

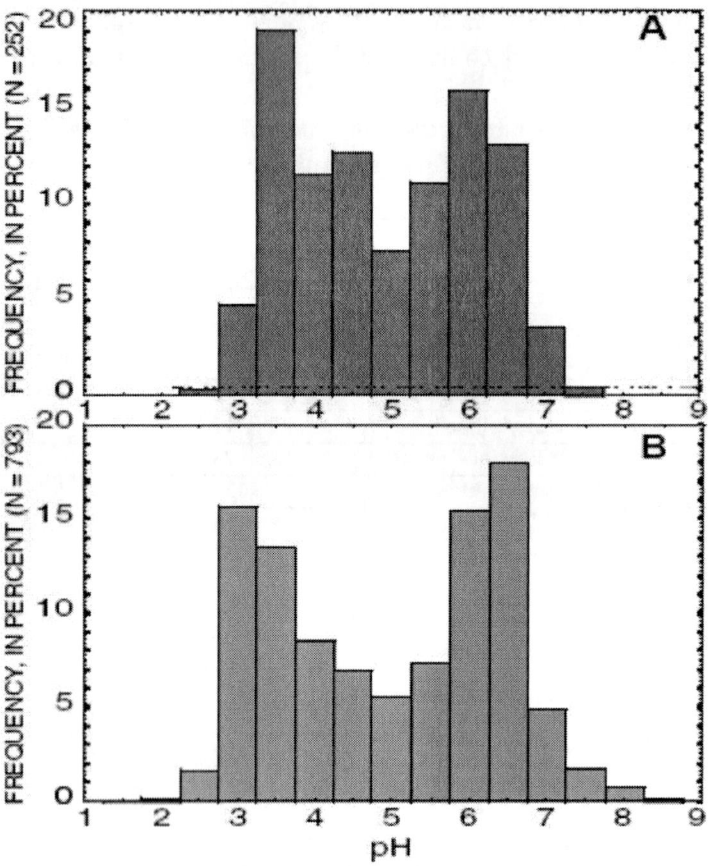

A) Data for 252 coal mine discharges in the anthracite coal field.
B) Data for 793 coal mine discharges in the bituminous coal field [Source Cravotta et al. (1999)].[40]

Figure 7-2. Frequency Distribution of the pH of Coal Mine Discharges in Pennsylvania.

Viability for Use

In light of the elevated levels of TDS (from 200 to more than 10,000 mg/L), hardness, and the variable nature of pH, mine pool water will have to be treated to some degree prior to use in a power plant cooling water system in all but the most unusual conditions. Treatment technologies for pH adjustment, removal of TDS, dissolved metal ions, and other constituents found in mine pool water are available.

An example of a treatment system of mine pool water that might be used for a closed-cycle cooling system makeup includes the following process units:

- Clarification to remove suspended solids.
- pH adjustment, horizontal precipitator, coagulation, and flocculation for metals removal.
- Multimedia filtration, ion exchange, and carbon adsorption if necessary to remove low TDS.

Any water that will be used for boiler feed water will require additional treatment. The following additional treatment units can produce high-quality boiler feed water:

- Cartridge filter units for additional sub-micron particle removal.
- RO and/or ion exchange for additional polishing.[42]

Specific treatment technologies for impaired waters are discussed in other Sections. The focus of studies on the use of mine pool waters in power plant cooling and operations is more closely related to evaluation of the physical feasibility of using this resource and developing methodologies to determine the level of treatment required to effectively use mine pool waters. Efforts in these areas are described below.

Two NETL studies are evaluating the viability of using mine pool water for cooling in coal-fired power plants. Both the University of Pittsburgh/Carnegie Mellon University study and the NMLRC/West Virginia University study are evaluating the impacts of geographic proximity and other physical characteristics on mine pool water viability. The University of Pittsburgh/Carnegie Mellon University study is also developing pilot-scale testing capabilities, while the NMLRC/West Virginia University study includes a computer-aided design component.

The overall objective of the University of Pittsburgh/Carnegie Mellon University study[45,46] is to assess the potential of using three different impaired waters (secondary treated municipal wastewater, passively treated coal mine drainage, and ash pond effluent) as cooling water in coal-based thermoelectric power generation. The study will assess geographic proximity, pretreatment requirements, available quantities, and regulatory and permitting issues that are relevant for application of these impaired waters in a cooling system. In addition, key design and operating parameters that would ensure successful use of mine pool waters without detrimental impact on the performance of the cooling system will be evaluated. Another important project objective is the development and demonstration of small pilot-scale cooling towers for side-byside evaluation of the use of impaired waters under different operating conditions. The pilot-scale cooling towers have been built and are pictured in Figure 7-3.[45]

In anticipation of the use of the pilot-scale towers with mine pool waters, preliminary bench testing of acid mine drainage (AMD) collected from a mine in western Pennsylvania revealed that this water could be suitable for use in cooling towers with the addition of 15 parts per millions (ppm) of PMA, 2 ppm of TTA, and 1 ppm of monochloramine to control scaling, corrosion, and biofouling, respectively.[45]

Following the bench-scale testing, the pilot-scale cooling towers were operated for 25 consecutive days using AMD from the same location as the bench-scale testing. Data recorded included water temperature at specific locations, airflow rate in the cooling tower, the conductivity of makeup and bulk water, makeup water volume, blowdown volume, water flowrate, and ambient conditions (weather, temperature, relative humidity).[46]

The impact of PMA on scaling was evaluated in one analysis. PMA was added to two towers at concentrations of 15 and 25 ppm. Scaling behavior was monitored based on mass gain of stainless steel coupons. This analysis generated counterintuitive observations in that higher concentrations of PMA resulted in more deposition on steel coupons. This behavior was explained by the observation that significant quantities of solids precipitated on the packing in the control tower (which did not receive any antiscalant material) and that the turbidity of the recirculating water in the control tower was close to that of the makeup water. These findings suggest that the solids formed in the control tower were easily separated from the liquid phase and removed from the system. This was evidenced by the significant mass of solids that accumulated in the bottom sump of the control tower. However, the presence of PMA in the test towers resulted in relatively high levels of suspended solids that were maintained in solution and given a chance to precipitate on the sampling coupons. Overall, PMA was not very effective at keeping the high dissolved solids of AMD in solution under the pilot testing conditions. The solids content of the AMD water at four cycles of concentration was very high, and inhibition of precipitation by PMA proved to be difficult.[46]

Corrosion of metal alloys in the pilot-scale cooling towers was monitored through weight loss measurement. The two test towers were operated with the addition of phosphate (target of 5 ppm) and 2 ppm of TTA. Study results indicated that aluminum is not a suitable material for cooling water systems using AMD as makeup water due to severe pitting corrosion. Mild steel and copper corrosion were successfully inhibited by the presence of TTA. Copper-nickel exhibited little corrosion with or without the presence of TTA. Phosphate-based inhibitors are not suitable for use in this system due to their low solubility in the recirculating cooling water.[46]

Biofouling and bioactivity were controlled by adding 1 to 2 ppm chloramine as Cl2 to each test tower. The chloramine was added into the system continuously from a stock solution. Experimental results showed that the continuous chloramine concentration used achieved very good control of both planktonic and sessile heterotrophic bacteria in the system, and that the common control objective for biofouling (10^4 CFU/mL) can be easily accomplished under test conditions.[46]

The NMLRC/West Virginia University study[47] is identifying mine water sources and quantifying the volume and water quality of mine water with the goal of determining whether it can be used for cooling at a specific power plant. Based upon preliminary analyses, there appears to be sufficient water available (3,000 gpm) from underground mines to meet the water needs of the proposed Beech Hollow power plant, a 300-MW facility burning waste coal. About half of this water need may be available in the Raccoon Creek watershed, including several creeks that are fed by mine discharges. Use of this water by the power plant would have a major impact on the restoration of Raccoon Creek between the JB-1 discharge (just under five miles southwest of the power plant site) and the Ohio River. Additional mine water is available in the Robinson Run and North Branch Robinson Run watersheds. These two tributaries to Chartiers Creek are severely impacted by AMD. Diversion of water to the power plant in these watersheds will reduce, but not eliminate, the acid load of the receiving streams. This illustrates a potentially important benefit associated with using mine drainage water in a power plant setting. If these impaired waters are diverted for beneficial use, treated, and then returned to surface or ground water, the negative impacts that could occur if the drainage waters flowed directly into surface waters can be avoided. A map showing the mine discharge locations relative to the power plant is presented in Figure 7-4.[42]

The wide distribution and generally small size of the mine discharges in the vicinity of the power plant will adversely affect the economics of complete mine water use at this site. However, this distribution will present a robust basis for developing the computer-based design aid that is one of the project deliverables. The design aid consists of an interactive Microsoft Excel spreadsheet with Visual Basic macros. Initial work on the design aid has focused on the design of the water piping system needed to transport the water from the mine discharge to the water treatment plant or power plant. Data analysis includes the length of pipe, static and dynamic head, pipe diameter, pipe cost, pump horsepower requirements, operations and maintenance (O&M) cost, and water temperature changes resulting from buried pipe flow.[47]

Figure 7-3. Pilot Scale Cooling Towers.[45]

While most of the alternatives involved using mine water in cooling towers, the West Virginia University study cited above also looked at the ability of abandoned coal mines to serve as a heat sink.[43] Under this scenario, cool water is extracted from a specific mine, run through the power plant, and then warm water is re-injected into the same mine or an adjacent mine. Flow and thermal modeling is required to establish design parameters for this type of cooling. Thermal modeling establishes the area of the mine needed to achieve the required cooling, while flow modeling produces average mine residence times that must be known to assure cooling performance. Two specific scenarios were evaluated and compared to a base case that assumed cooling water was supplied from a conventional source (e.g., surface water). The first scenario involved injecting hot water from the power plant into one mine and extracting cooled water from an adjacent mine. The median underground travel time between mines was approximately 206 days. The capital cost associated with this scenario was 75 percent of the base case due to elimination of the cooling tower. However, operating costs were 119 percent of the base case due to increased water pumping and treatment requirements. The overall cost-effectiveness of this scenario will be site-specific. A second scenario involved multiple extraction and injection steps to try to use cooler water in the power plant. This involved injecting hot water from the power plant into one mine, extracting the cooled water from an adjacent mine, reinjecting the cooled water into a third mine, and extracting the cooler water from the upper part of the third mine. This scenario had capital cost that were 80 percent of the base case with operating costs that were 193 percent of the base case.[45]

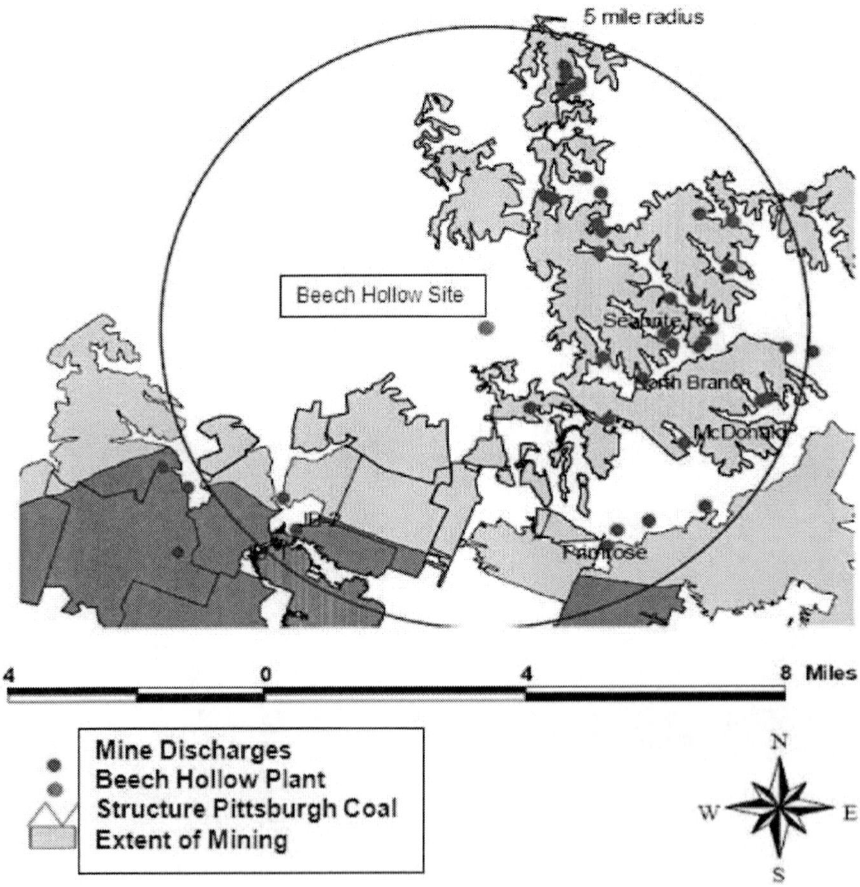

Figure 7-4. Mine Discharges in the Vicinity of the Proposed Beech Hollow Power Plant.[47]

Other Issues

Gillette and Veil[48] examined the non-technical issues associated with the use of mine pool water. They noted that there are technical, economic, legal, regulatory, environmental, and public perception issues that must be identified and resolved. Specifically, they noted that regulatory and legal issues must be resolved to make widespread use of mine pool water resources feasible. Any company using mine pool water would need to meet the provisions of their NPDES permits, as well as any other discharge requirements that might be imposed at any level of government. Measures needed to meet regulatory requirements on water discharge and their associated costs are perhaps the leading barriers to wider use of mine pool water. Many regulatory agencies at all levels of government have not traditionally addressed these waters in a way that reflects their potential use as a valuable resource.[48]

Technical issues to be resolved include methods of effectively collecting, treating, and distributing the water and its long-term sustainability. The issue of subsidence following the withdrawal of mine pool waters could also be an important consideration.[48]

While each of these issues represents a potential barrier to developing mine pool water resources, the cost of using these waters is considered the greatest general barrier that must be overcome. It is also the barrier against which government incentives are likely to be most effective. The overall cost of using mine pool water includes the following components:

- **Collection** – Mine pool waters might need to be collected from a variety of locations in order to meet volumetric rate requirements.
- **Treatment** – The water may require treatment to prevent scaling or corrosion problems.
- **Transportation/Distribution** – Depending on the relative locations of the source, use, and discharge, there may be costs associated with transporting and/or distributing mine pool water.
- **Disposal/Discharge** – Costs may be incurred depending on the use of the water and the physical conditions with respect to discharge points (rivers, lakes, ground water).[48]

Gillette and Veil listed possible incentives for using mine pool waters such as direct grants, tax/royalty subsidies or reductions, reduced water costs to the user, assured markets, and regulatory relief. They then constructed six scenarios for mine pool waters and presented examples of such incentives. Those scenarios would be examined in a further study to assess both quantitative and qualitative metrics for potential incentives.[48]

Yet another issue related to the use of mine pool waters was addressed by Watzlaf and Ackman (2006). They point out that use of mine pool water in geothermal heat pumps could be an extremely cost-effective method to provide residential and/or industrial heating and cooling. This is particularly true where the mine water is already being pumped and treated. Operational costs are much lower than conventional heating and cooling. Costs per unit of heat are only 33 percent, 34 percent, and 21 percent of the costs associated with using fuel oil, natural gas, or propane, respectively, for heating and cooling. They point out that the amount of water currently being discharged for mines in the Pittsburgh coal seam could potentially be used to heat and cool up to 3.74 million square meters of interior space, roughly equivalent to

20,000 homes.[49] This type of system has been deployed successfully at abandoned mine sites in Canada, the United States, and Europe.[49]

Conclusion

Mine pool waters represent a significant volume of available water that may be transported to nearby power plants for use in cooling systems. While the volume of mine pool waters is large, there are questions about the sustainability of these waters as supplies to cooling water systems. However, studies have shown that there are indeed mine pools that can provide sustainable waters to power plants and supplement other water supplies. Mine pool waters are generally low in pH and have high TDS, so they likely require treatment prior to use in cooling systems.

The University of Pittsburgh/Carnegie Mellon University study currently underway is considering factors such as geographic proximity, pretreatment requirements, available quantities, and regulatory and permitting issues that are relevant for application of these impaired waters in a cooling system. In addition, key design and operating parameters that would ensure successful use of mine pool waters without detrimental impact on the performance of the cooling system will be evaluated. Another important project objective is the development and demonstration of small, pilot-scale cooling towers for side-by-side evaluation of the use of impaired waters under different operating conditions.

Testing conducted to date in these pilot-scale cooling towers indicates that using AMD as makeup water leads to scaling issues that are not ameliorated by the addition of PMA. Corrosion analysis indicated that aluminum should not be used in the presence of AMD, but that mild steel and copper corrosion is inhibited by the presence of TTA, and copper-nickel corrosion is minimal with or without TTA. Biofouling was effectively controlled by chloramine treatment.

The NMLRC of West Virginia University is identifying mine pool water sources and quantifying the volume and water quality of mine water with the goal of determining whether it can be used for cooling water at the proposed Beach Hollow power plant. Mine drainages from several mines surrounding the power plant site have been evaluated in terms of both flow rates and water quality to determine their viability for use in the plant. One of the positive outcomes of using the mine drainage water in the power plant is that impaired water that is currently having negative impacts on local streams would be diverted from those surface waters, improving their water quality. Another product of this project is the development of a computerized design tool that can be used to evaluate the viability of using mine pool water for cooling in other specific applications.

A second West Virginia University study has evaluated the use of abandoned coal mines as a wide area heat sink. This involves extraction of cool water from a specific mine that is used for power plant cooling and then the warmed water is re-injected into the same mine or an adjacent mine. Flow and thermal modeling is required to establish design parameters for this type of cooling. Based upon two scenarios evaluated, capital costs are lower for this type of operation than more traditional cooling methods, but operating costs are higher.

Some regulatory issues also need to be addressed if mine pool waters are to be considered for use at power plants. Regulatory agencies at all levels of government have not traditionally addressed these waters in a way that reflects their potential use as a valuable resource. The government could play a key role in the future use of these resources through incentive programs that reduce the cost and associated risks inherent in using these non-traditional waters.

Finally, the use of mine pool waters in geothermal heat pumps could be a cost-effective method to provide residential and/or industrial heating and cooling. Costs associated with geothermal heating and cooling are significantly lower than those associated with conventional methods of heating and cooling (e.g., natural gas and fuel oil). Analyses indicate that water currently being discharged just by the Pittsburgh coal seam could be used to heat and cool up to 20,000 homes.

SECTION 8. PRODUCED WATER FROM CARBON DIOXIDE STORAGE IN SALINE FORMATIONS

This Section covers Existing Plants Program-funded work on water produced from saline geologic formations when CO2 is sequestered (stored) in these formations thousands of feet below the surface. Geologic sequestration is an option that would reduce CO2 emissions to the atmosphere and help mitigate climate change. Section 6 considers the produced water resulting from geologic sequestration in fossil fuel beds for the purposes of recovering oil and gas, but the Nation's saline formations offer substantially more CO2 storage capacity than the fossil fuel beds.

This Section reflects work that has been, or is currently being done, in several different NETL studies, including:

- "Use of Produced Water in Recirculating Cooling Systems at Power Generating Facilities," (NT41906) led by EPRI at the SJGS that examined treatment options for oil and gas well produced waters.
- Year I of "Study of the Use of Saline Aquifers for Combined Thermoelectric Power Plant Water Needs and Carbon Sequestration at a Regional Scale," (07-013812) completed by SNL in June 2008 that considered the geochemistry of saline formations and developed cost estimates for providing cooling tower makeup water.
- Year II of "Study of the Use of Saline Aquifers for Combined Thermoelectric Power Plant Water Needs and Carbon Sequestration at a Regional Scale," (07-013812) led by SNL that will expand the assessment.
- "Thermoelectric Power Plant Water Demands Using Alternative Water Supplies: Power Demand Options in Regions of Water Stress and Future Carbon Management," (FWP-08-014053) examining additional saline formations across the Nation and led by SNL.
- "Technology to Facilitate the Use of Impaired Waters in Cooling Towers," (NT0005961) coupling electrodialysis reversal (EDR) to lower TDS and a silica-specific ligand functionalized core material to remove silica.

The water extracted from saline formations will have been forced by high pressure CO_2 injection to extraction wells at other locations within the formation. An advantage to using saline formations may be that the available volumes are large enough to provide a considerable amount of water for an extended period of time. The SNL study[50] indicates that each metric ton of CO_2 injected into a saline formation may displace roughly 317 gallons of water. Because considerable uncertainty exists with these underground formations, capacity estimates are conservative.

Additional research into using saline formations for CO_2 storage is being done by the NETLfunded work of NATCARB.[51] The NATCARB project is exploring the geological sequestration of carbon by combining regional databases on geology and emissions into an interactive mapping system. Links at http://www.natcarb.org/Atlas/sinks.html can help interested parties understand the geospatial extents of these saline formations.

Environmental Concerns for Untreated Waters

According to the SNL study, the composition of the waters extracted from saline formations will change in later years as compressed CO_2 dissolves in the water and reacts chemically with the mineral formations. Carbon dioxide in water produces carbonic acid, a weak acid. When acid exceeds the buffering capacity of the waters, the pH will decrease. The pH of the water in contact with the CO_2 is expected to initially drop from near 8.0 to a range of 3.5 to 5, depending on the surrounding minerals. More minerals will dissolve in the acidic environment, but slow processes will eventually precipitate carbonate minerals again.

The goal would be to only extract water that had not been in contact with the injected CO2, so changes in acidity are not of concern. However, the disposal of saline formation waters could still present permitting challenges for sequestration operations with regard to discharge limits on TDS. Sequestration activities in saline formations will likely be restricted to formations with TDS greater than 10,000 mg/L to protect potential drinking water sources; so saline formation sequestration produced waters will have TDS greater than 10,000 mg/L.

Access and Collection

The water extracted from saline formations will be produced at extraction wells located in proximity of the injection wells. In the case of SJGS, an earlier study focused on the Fruitland Formation suitability for water extraction and carbon sequestration.[52] Carbon sequestration in the Fruitland Formation was accompanied by oil/gas recovery, and the project was discussed in Section 6. Wells in the Fruitland formation are located 32 miles from the power plant, so the water required delivery from the extraction wells to the plant. The investigation examined trucking, utilizing abandoned gas pipelines, and constructing new pipelines as options for transporting the water to the power plant.

The 2008 report assumed that the Morrison Formation would be used and that water would need to be pumped from the well to SJGS a distance of three miles. Five other saline formations located within 35 miles of SJGS and the Four Corners Plant were also evaluated in SNL's report. It was assumed that the desalination process would occur at the power plants, but this study did not examine different transport processes. Readers interested in the costing of the access and collection process from extraction wells should refer to the Access and Collection discussion in Section 6.

Non-Traditional Water Quality Assessment

The SNL study devoted a Section to the prediction of the chemical composition expected from the various saline formations. The introduction of CO_2 into the formations will alter the chemical equilibrium state, but the kinetics are often slow. Therefore, the SNL study used box models (including TOUGH-2) to predict the aqueous concentrations initially, after 100 years, after 350 years, and when the equilibrium is balanced. This method was applied for six different local formations; the ranges in chemical composition after CO_2 is added over time are shown in the following table. Table 8-1 compares the concentrations for the nearby Morrison Formation with the limits for all six formations.

Table 8-1. Box Model Predictions of Chemical Composition for CO_2 Storage in Saline Formations

Water Chemistry Composition	Minimum Concentration in Morrison Formation (ppm)	Maximum Concentration in Morrison Formation (ppm)	Minimum Concentration in All Formations (ppm)	Maximum Concentration in All Formations (ppm)
Aluminum	0.004	0.008	0.001	0.014
Calcium	278	1767	7.1	2056
Carbon	22,220	30,700	22,220	31,360
Chlorine	62.8	65.1	62.8	3810
Iron	0	29.7	0	32.9
Magnesium	43.6	786	6.2	1709
Potassium	8.9	5845	8.9	7511
Silicon	2.4	21.8	2.3	24.5
Sodium	194.2	1393	184	4968
Sulfate	963	3351	1.24	4899
pH range	3.49	5.31	3.45	5.31

Considering all six local formations, the SNL study summarized that initially high silica and iron concentrations would present desalination difficulties, as would the high calcium content. Acidic pH levels would last for centuries. However, when full equilibrium is reached (far beyond 350 years), the pH would be expected to be only slightly acidic, potassium and magnesium concentrations would rise, and calcium and sodium concentrations would equilibrate to low levels. Because the produced waters are expected to consist of water ahead of the CO_2 plumes, the initial chemistry of the waters is of interest.

The salinity across these formations is not always uniform, and the shallow components of the formations may be less salty than the deeper components. At the October 2008 Existing Plants Water Projects Meeting, it was suggested that saline formations located beneath western salt domes may also have concerns with arsenic and selenium in the produced waters.

An expanded study by SNL[53] has sought to describe additional produced waters within saline formations in other water-stressed areas of the country. The deep saline formations initially investigated in the Southeast had TDS far greater than the 20,000 mg/L selection criteria initially assumed for feasible treatment. With the criteria stating that the formations are to be located at least 2,500 feet below the surface and the TDS be between 10,000 and

20,000 mg/L, only one of the four sites initially investigated in the Southeast was suitable. The Black Warrior Basin Coal Test Site is located in northern Alabama, but SNL is also investigating potential sites in Florida, Georgia, and Texas. As part of this study, SNL is developing a tool to guide other investigations of suitable carbon sequestration sites for produced waters that could be treated for power plant use.

Treatment Technology Descriptions

For the initial run of their model, SNL investigated the site specifics of the Morrison Formation, chosen for its geochemistry, physical location, and for desalination purposes. The brackish waters from this formation could not be used as cooling tower makeup without treatment (chlorides must be below 1,000 mg/L). A simple spreadsheet analysis was conducted to study the costs and yields of four desalination options:

- Desalination and gathering equipment only; no concentrate disposal.
- Desalination and gathering equipment only; 59.5 acre evaporation ponds for concentrate disposal.
- Desalination and gathering equipment only; 3,000-ft pipeline and injection well for concentrate disposal.
- Desalination and gathering equipment with HERO process and a BC retrofit.

SNL's study chose the final option with an estimated cost of $5.32 per 1,000 gallons of produced water for these site-specific conditions. The treatment technology was estimated at approximately five percent of the base costs; in comparison, the carbon capture and sequestration (CCS) process represented an additional 100 percent to the base costs.

The HERO+BC treatment was chosen over other options based on the findings from a previous NETL/EPRI study done at SJGS.[52] The earlier study is described in Section 6 that details treatment of produced water from oil and gas wells. The HERO process pre-treats the water to remove scale-forming minerals (e.g., calcium) and other ions associated with high salinity (e.g., chloride). The subsequent BC separates the reject water from the HERO process into mineral-rich and mineral-lean streams. The recovered mineral-lean water from the BC and the HERO permeate would be fed to the cooling towers.

Like conventional RO techniques, the HERO process uses standard, spiral-wound polyamide membranes. It differs from conventional techniques because it operates at high pH values (9.5- 10.5) after pre-softening to remove hardness. At pH levels above 10, soluble silica (H4SiO4) dissociates and does not form scale on the membranes.

In addition, pretreatments will be necessary in order to ensure proper operation of RO systems. These include precipitation softening or scale inhibitors to prevent mineral scale from fouling membranes, filtration to remove particulate matter, periodic membrane cleaning with non-oxidizing biocides to remove biofilms, a combination of gravity separation, air flotation, and filtration to remove oil, and neutralization of charged colloidal matter with polymers. The saline formation waters may not require all of these pretreatments (e.g., oil filtration), and the high pH of the HERO system dissolves organic fouling and certain oil constituents to minimize fouling by these mechanisms. An illustration of the pretreatments and the HERO system (designed for oil and gas well waters) is shown in Figure 8-1.[52]

Following the exit from the HERO system, RO reject stream would be sent to the BC for further water recovery. The BC would operate in the un-seeded mode at high pH (10-11), eliminating the need for acid, antiscalant, and calcium chloride.

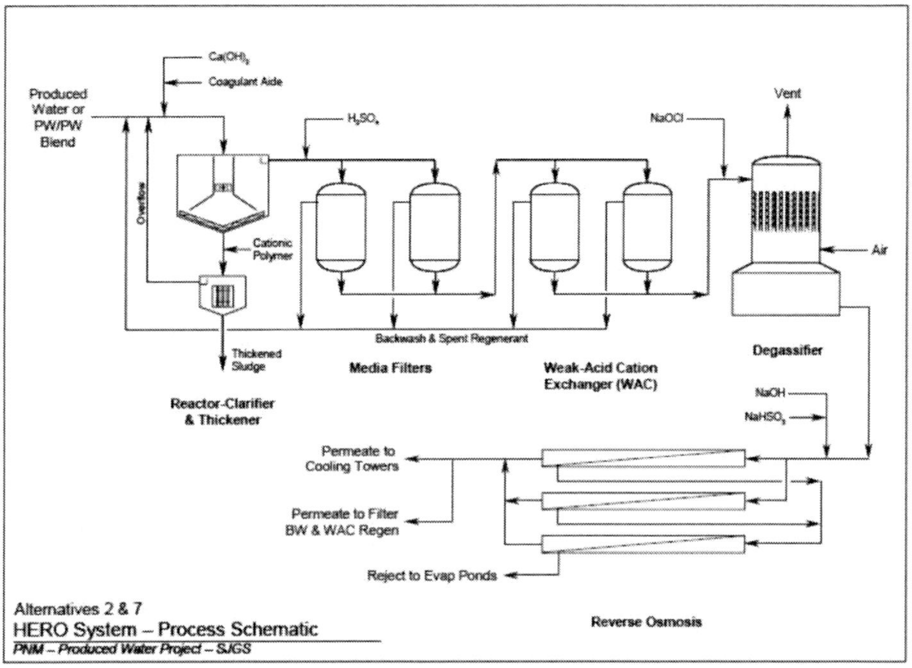

Figure 8-1. Illustration of Pretreatments and HERO System for Produced Waters.[52]

Another process for silica removal is beginning development at GE Global Research. Impaired waters with silica concentrations of 100 ppm will first be treated through EDR, an electrochemical separation process that removes ions and other charged species from fluids.[54] Direct current voltage transfers the ions through ion exchange membranes to desalinate the stream. The GE EDR technology process has been marketed for municipal drinking water, industrial process water, and wastewater reuse, and it includes the following features:

- High water recovery rate design (and the associated reduced waste discharges)
- Membranes that tolerate moderate particulate levels
- Long membrane life
- Chlorine-resistant membranes
- Low electricity consumption
- High silica tolerance

GE Global Research proposes that the self-cleaning EDR membranes will reduce the TDS from 2,500 mg/L to the range of 25-100 mg/L.[55] Silica remediation will then be performed on the effluent stream to reduce the silica concentrations by 90 percent using ligand functionalized core materials (LFCM) developed with silica-specific ligands. The bound silica will then be separated, and the LFCM will be recycled after the silica is removed.

Performance Goals

For produced water from the Morrison saline formation, the HERO process is designed to operate with an 83 percent efficiency, producing 0.83 gallons of water for each gallon of feed. The concentrate stream from the HERO process would be treated by the BC to recover 73 percent of that reject. Total solids from the BC would be limited to the 200,000 to 250,000 mg/L range.

The SNL study has estimated the costs associated with this process based on the assumptions shown in Table 8-2. Based on these assumptions, SNL estimated the total treatment cost for the Morrison Formation at $5.32 per 1,000 gallons of treated water (including ponds and groundwater pumping).

The work at GE Global Research is currently in Phase I (Material Selection and Synthesis).[55] The tasks within this phase are aimed at synthesizing and testing LFCMs for silica and developing preliminary cost estimates. The goal in this phase is to achieve 90 percent silica removal. Later phases will include demonstration of a recycle protocol at the benchtop level, scale-up of the LFCM synthesis, pilot plant design and demonstration, and finalization of cost and process models.

Other Issues

Power plants considering CCS must identify additional water sources to be used in the carbon capture process (Table 1-1). A DOE/NETL report[56] examined the effects of carbon capture technologies on three scenarios that estimated PC plants required 143-147 BGD of water withdrawal. The carbon capture process that was constructed on the coal-fired fleet with existing scrubbers demanded an additional 0.7-4.1 BGD of water withdrawal. A 2008 analysis[57] shows lower water withdrawal and consumption rates for all future years compared to the 2007 analysis largely because the future thermoelectric capacity estimates decreased in the 2008 analysis.

On July 25, 2008, the U.S. EPA proposed a rule designed to outline the requirements necessary to develop wells for geologic sequestration of CO_2.[58] Previous injection well requirements under EPA's underground injection control (UIC) program had established five different categories for wells. The issues associated with the geologic sequestration of CO_2 for storage purposes were sufficiently different than previous well activities, so EPA proposed a Class VI well with its own requirements. These requirements are designed to ensure safe injection of CO_2 while protecting potential drinking water aquifers.

The salts produced during this treatment require disposal, but the SNL study does not make recommendations about the appropriate techniques. Because the produced waters containing the same quantities of salts would also require disposal, this aspect was not considered in the cost estimates.

Year II of SNL's study of using brackish water from saline formations expands the assessment to consider additional factors. It will characterize the CO_2 sequestration potential within select formations, consider new desalination technologies and potential uses of waste heat for reducing desalination costs, and incorporate considerations of the costs associated with CO_2 sequestration within these formations. The study will continue to focus on SJGS but will create a framework that allows planners to consider the development costs of collocation of power plants near saline formations to both utilize the carbon sequestration potential and the available brackish water. Year II of the study will also consider using the brackish water in systems other than cooling towers.

In addition, another study by SNL (Project FWP-08-014053) has been investigating candidate sites in other parts of the United States where potential geologic sequestration formations may provide water. That study is described further in Section 10.

Table 8-2. Summary of Desalination Calculations and Assumptions for CO_2 Storage Waters

	Base Case	Value
Water TDS (mg/L)	Used Morrison formation	6,000
Design Flow rate (gpm)	Based on 24/7 pumping of brackish aquifer and % CO2 capture	1,807
Design Flow Rate (MGD)	Used 2.0 MGD desalination output (treated water)	
Design Annual Flow (Mgal/yr)	Based on 0.85 plant capacity factor (USBR recommendation)	400
Electrical cost ($/kwh)		0.1
Pipeline distance from brackish well to desal plant (mi)	Based on radial distance, Morrison formation	3
Well Depth (ft)	Based on Morrison formation	4,725
Capital Costs:		
Pump & Pipe – Produced Water Gathering Capital	Used USBR desalting Handbook Figures 9-18 a	$2,000/ft
Piping from gathering station to desal plant	Used USBR desalting Handbook Figures 9-11 a	$126,810/mi
Concentrate Disposal pipeline & well	Used USBR desalting Handbook Figures 9-11 & 9- 13 a	
Evaporation Ponds	Used USBR desalting Handbook Figures 9-12 a	
Desalination Total Construction Cost	Used USBR desalting Handbook Figures 9-7 a	2,000 mg/L TDS for options A-C NETL/EPRI (2006) Value for option D
O&M		
Labor for (2) MGD	Used USBR desalting Handbook Figures 9-37 a	
Electrical BWRO (for 6,000 mg/L TDS, 2 MGD)	Used USBR desalting Handbook Figures 7-8 & 9- 45[a]	
Electrical GW pumping (for 1807 gpm/2 MGD)	Used equations to estimate pump power; see Appendix X	
Membrane Replacement	Used USBR desalting Handbook a	$0.08/1000 gal plant capacity
Chemicals (used surface water)	Used USBR desalting Handbook Figures 9-41 a	
Other Maintenance	Used USBR desalting Handbook Figures 9-18 a	1.5% of capital

[a] Desalting Handbook for Planners, third edition, United States Department of Interior Bureau of Reclamation, July 2003.

Conclusion

The first year of the SNL study has scoped the options associated with the use of waters produced from saline formations when accompanied by carbon sequestration in the formation. The extraction, transport, and treatment of the waters were examined in order to estimate the costs of using these waters for cooling tower makeup at SJGS. Site-specific costs were estimated at $5.32 per 1,000 gallons of treated water (including ponds and groundwater pumping).

The treatment costs were estimated for a single nearby formation (Morrison Formation) that could supply sufficient rates of brackish water to justify use as a makeup system. After pretreatment, the HERO system should recover 83 percent of the water, and the subsequent BC an additional 73 percent of the reject.

Year II of SNL's study will characterize the CO_2 sequestration potential within select formations, consider new desalination technologies and potential uses of waste heat for reducing desalination costs, and incorporate considerations of the costs associated with CO_2 sequestration within these formations. The study will continue to focus on SJGS but will create a framework that allows planners to consider development costs. Year II of the study will also consider using the brackish water in systems other than cooling towers. An additional ongoing study by SNL is examining additional potential geologic sequestration formations around the United States and evaluating their potential for using these non-traditional waters.

A new study is underway at GE Global Research to treat impaired waters with high TDS and silica concentrations. EDR techniques will first lower the TDS, and then a silica-specific ligand will be introduced to separate silica from the water stream. The ligand will be bound as a ligand functionalized core material, separated from the water stream, and recycled after the silica has been removed.

Section 9. Recovered Plant Discharges

This Section covers Existing Plants Program-funded work on water that normally would be discharged from the power plant, before or after treatment. To date, the work in this area has focused on discharges associated with ash pond effluent. The ash pond waters consist predominantly of waters contacting ash in the process stream (e.g., sluice) but also include surface water runoff from precipitation events. The ash ponds are primarily utilized for settling solids and provide only minimal treatment for inorganics and organics.

This Section reflects work that has been, or is currently being done, in two NETL studies:

- "An Innovative System for the Efficient and Effective Treatment of Non-Traditional Waters for Reuse in Thermoelectric Power Generation," (NT42535) led by Clemson University that tested how constructed wetlands can treat ash basin waters.
- "Reuse of Treated Internal or External Wastewaters in the Cooling Systems of Coal-based Thermoelectric Power Plants," (NT42722) led by the University of Pittsburgh that tests treated water performance in pilot-scale cooling towers.

Environmental Concerns for Untreated Waters

National attention has turned to the regulation of coal ash impoundments in recent months, and U.S. Representative Nick Rahall, chairman of the House Natural Resources Committee, proposed a bill (Coal Ash Reclamation and Environmental Safety Act of 2009) that would impose uniform design, engineering, and performance standards on coal ash impoundments.[59] Other members of Congress have also pushed for EPA to take actions to regulate operation of coal ash impoundments.

The Clemson University work surveyed chemical contaminants of concern in various ash basin waters, as indicated in Table 9-1.[60] In addition, the University of Pittsburgh noted the typical ash pond water quality range reported by 10 coal-fired power plants operated by Tennessee Valley Authority, which are also shown in Table 9-1.[61]

Although ash pond waters are not discharged from the plant, researchers compared them to NPDES and EPA-recommended water quality criteria (WQC) to represent their relative toxicities. Arsenic concentrations in ash pond waters exceed both NPDES permit levels of 0.293 mg/L and the WQC of 340 jig/L. The WQC for cadmium of 20 jig/L is lower than the concentrations in several basins listed in Table 9-1. NPDES permits restrict chromium concentrations to 0.2 mg/L, and the WQC recommends 570 jig/L and 16 jig/L as the upper limits for trivalent and hexavalent chromium. As indicated in Table 9-1, the Savannah River site exceeds these concentrations. For copper, both the NPDES permit limits (0.04-1.0 mg/L) and the EPA WQC (13 jig/L) were lower than many of the measured concentrations. The WQC for mercury is 1.4 jig/L; therefore, this element is a constituent of concern based on the measured concentrations. The WQC for selenium is only 5 jig/L, but most of the concentrations listed in Table 9-1 exceed this value. The EPA WQC for zinc recommends that the concentration not exceed 120 jig/L, but the concentrations in ash basin waters range from 0.011 to 8.1 mg/L.

Non-Traditional Water Quality Assessment

Studies also address the suitability of the recovered plant discharge waters for use at the power plants. Clemson reports that ash basin waters have low ionic strength and thus do not promote the formation of chemical scale. Because some constituents (e.g., mercury, arsenic, and zinc) pose risks to living organisms, Clemson lists biofouling as an issue of minor concern for the use of ash basin waters. In addition, scaling and biofouling were monitored in simulated ash pond waters. The principal issue of concern for ash basin waters is corrosion potential, and it was examined in the Clemson study.

The University of Pittsburgh study examined ash pond water from the Reliant Energy power plant.[62] In bench-scale tests, they found that ash pond water was less likely to cause biofouling than using secondary treated municipal wastewater. Their results show that both the planktonic and sessile heterotrophic plate counts for the untreated ash pond waters were close to, or only slightly exceeded, the target criterion.

The University of Pittsburgh examined the scaling behavior of actual and synthetic ash pond waters. The scaling results caused by raw water and water after four cycles of concentration (evaporated down to one quarter of the volume) were examined. An unexpected result was that the actual raw water seemed to produce more scaling deposits than the concentrated raw water in the bench-scale tests. The inversion in results was explained by both deposition of minerals during the concentration phase and the presence of deposits introduced to the raw water by mild steel corrosion coupons. Increased scaling in the synthetic ash pond waters was associated with higher calcium levels, but there was little influence from magnesium.

Table 9-1. Reported Chemical Contaminant Concentrations in Untreated Ash Basin Waters

Site	Contaminant	Concentrations (mg/L)
Savannah River, SC[a]	Arsenic	4.08±9.70
	Cadmium	0.1083±0.061
	Chromium	0.324±0.382
	Copper	1.003±0.901
	Mercury	0.0267±0.007
	Selenium	0.79±1.52
	Zinc	3.27±2.839
Virginia site[a]	Arsenic	0.450±0.424
	Cadmium	0.17 (n=1)
	Mercury	0.025±0.007
	Zinc	2.67 (n=1)
New Mexico site[a]	Arsenic	0.030
	Cadmium	0.001
	Chromium	0.0025±0.0007
	Copper	0.0025±0.0007
	Selenium	0.058±0.002
	Zinc	0.510±0.098
North Carolina site[a]	Arsenic	0.039±0.061
	Selenium	0.0158±0.0167
	Zinc	0.030±0.023
Tennessee Valley Authority[b]	Aluminum	0.1—8.8
	Arsenic	<0.005—0.180
	Beryllium	<0.01—0.002
	Cadmium	<0.001—0.052
	Chromium	<0.005—0.17
	Copper	<0.01—0.45
	Cyanide	<0.01—0.02
	Lead	<0.01—0.2
	Mercury	<0.0002—0.3
	Selenium	<0.001—0.14
	Silver	<0.01—0.03
	Zinc	<0.01—2.7

[a] reported by Clemson University
[b] reported by the University of Pittsburgh

Treatment Technology Descriptions

The University of Pittsburgh study has measured the effects of different chemical treatments on ash pond waters for use in power plants through bench-scale testing but has not had an opportunity for pilot-scale testing. Their efforts have focused on scale control, chlorine demand, and biofouling prevention.

The University of Pittsburgh tested scale control additives in raw and simulated ash pond waters, but the tests in raw waters were not conclusive. In bench-scale testing, the synthetic waters were tested and compared with waters dosed with polyacrylic acid (PAA), PMA, and PBTC individually, as well as a 2:1 dosage ratio of PMA to PBTC.

Bacterial control in cooling water often depends on maintenance of a free chlorine residual in the range from 0.5 to 1 ppm Cl2. The ash pond waters were tested to determine the chlorine dosage necessary to maintain these concentrations. Biofouling tests (both planktonic and sessile heterotrophic plate counts) were also conducted on the untreated raw ash pond waters to determine biofouling rates.

In the CWTS designed at Clemson University, water passed through one of two reactor series, but both series consisted of four similarly constructed freshwater wetlands reactors (Figure 9-1). The sequential order runs from a moderately reducing wetlands reactor (redox potential Eh from -100 to 100 mV) followed by two smaller, more strongly reducing reactors (Eh less than -150 mV) and finally an oxidizing wetlands reactor (Eh from -100 to 250 mV). The reducing reactors include 30 cm of river sand hydrosoil and *Schoenoplectus californicus* (California bulrush) planted at realistic field densities. The oxidizing reactors include a granite cascade, 30 cm of river sand hydrosoil, and *Typha angustifolia* (narrowleaf cattail).

Table 9-2 shows characteristics of the reactors prior to introduction of the simulated ash basin waters. Ash basin water was simulated by adding high-purity salts of arsenic, chromium, mercury, selenium, and zinc to the reservoir. The retention time in the reactor series was approximately five days.

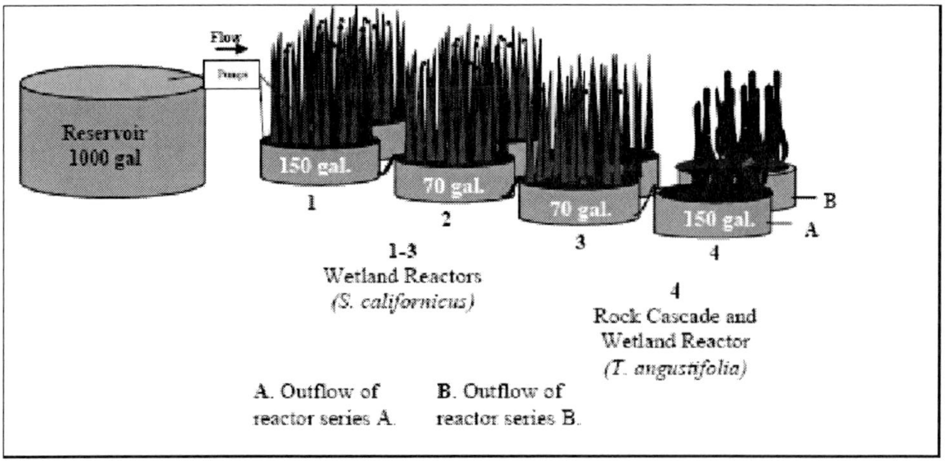

Figure 9-1. Illustration of CWTS for Ash Basin Waters.

Table 9-2. Parameters Indicating CWTS Readiness for Treatment of Ash Basin Waters

Series and Reactor	A1	B1	A2	B2	A3	B3	A4	B4
Eh (mV)	-76	-67	-172	-174	-182	-167	189	-17
Sediment Organic Matter (%)	0.10	0.12	0.42	0.24	0.36	0.25	0.19	0.18
Plant Density (#/m^2)	272	436	340	432	336	384	152	120

Performance Goals

The goal for the University of Pittsburgh study is to demonstrate that ash pond waters can be chemically treated to perform as cooling tower makeup waters in pilot-scale testing. The efforts focus on scale control through the addition of PAA, PMA, PTBC, or a combination and biofouling prevention through chlorine addition.[63] Bench-scale testing in a recirculating system showed that significant biocide and inhibitor concentrations were not necessary for biofouling prevention in the ash pond waters at both one and four cycles of concentration. When the free chlorine levels rose above 0.2 ppm, the planktonic bacteria decreased to non-detectable levels. However, free chlorine concentrations at 0.5 ppm increased the copper corrosion rate six times without inhibitors and four times with an inhibitor mixture. Investigators recommended that free chlorine levels remain at or below 0.5 ppm to achieve biofouling control without significant corrosion.

The University of Pittsburgh conducted bench-scale testing of scaling control methods using simulated ash pond waters in order to maintain better controls than could be achieved with the raw waters. The simulated waters representing four cycles of concentration were composed of Ca^{2+} (177 mg/L), Mg^{2+} (39.6 mg/L), Na^{+} (217 mg/L), Fe^{3+} (15.6 mg/L), $SO4^{2-}$ (392 mg/L), $HCO3^{-}$ (312 mg/L), and Cl^{-} (312 mg/L). The bench-scale testing (pictured in Figure 9-2) showed that PAA had the least impact on scale formation beyond four days, and PMA and PBTC showed similar performance by reducing scaling deposits roughly four times. When PMA and PBTC were dosed at 10 ppm and 5 ppm in the same test, the antiscaling effectiveness improved, but not significantly.

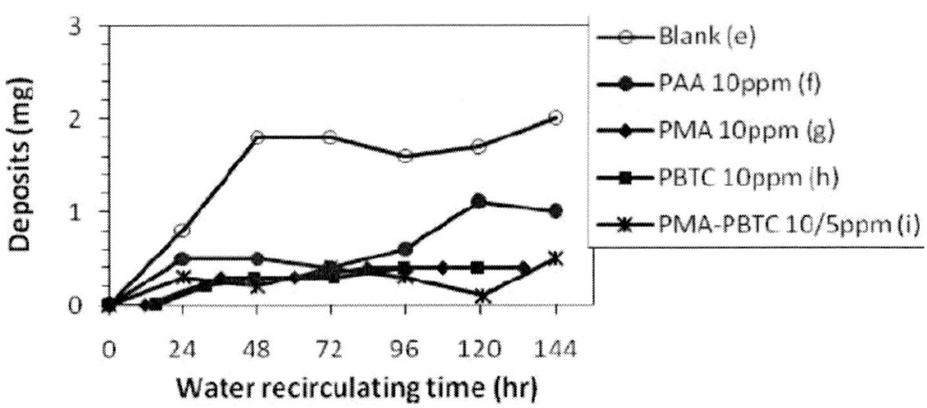

Figure 9-2. Scaling Behavior of Synthetic Ash Pond Effluent in Bench-Scale Tests: Effectiveness of Different Antiscalants at Four Cycles of Concentration.

Performance Results

These two studies have addressed different water treatment for different purposes. The Clemson University study investigated using wetlands for the removal of toxic constituents for discharge. The University of Pittsburgh study researched using chemical treatment to control the corrosion, scaling, and biofouling characteristics for reuse in power plant cooling towers.

The Clemson CWTS showed decreases in concentrations of arsenic, chromium, mercury, selenium, and zinc below the detection limits (sample shown in Figure 9-3), and the detection limits will be lowered for future testing. In addition, the conductivity, dissolved oxygen, alkalinity, and hardness remained fairly constant during the three experimental periods. Toxicity tests showed that survival and reproduction rates were higher in the treated waters than the untreated waters.[64]

The Clemson investigators also found that scaling and biofouling potential decreased significantly through both reactor series compared with the original detention basin (Figure 9-4). However, Figure 9-4 shows that corrosion potential increased significantly.

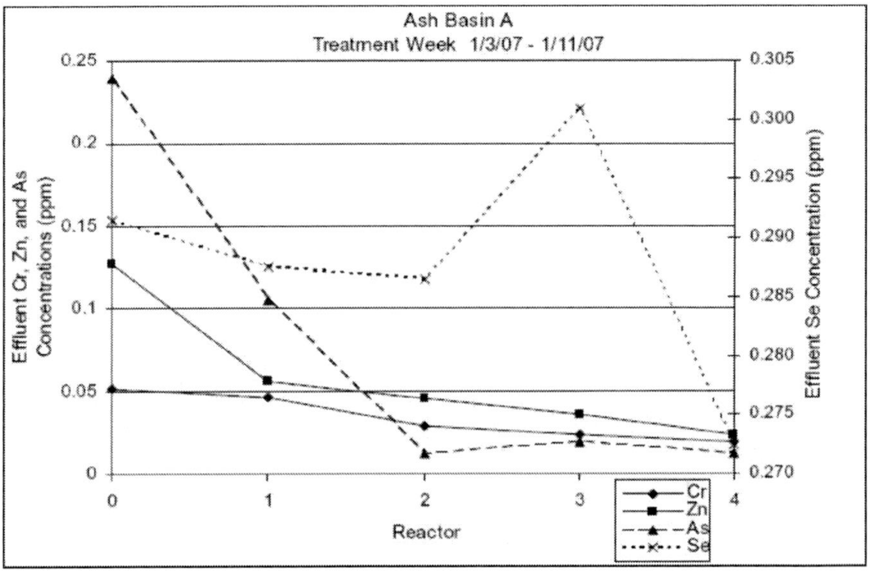

Figure 9-3. Sample Chart Showing Element Removal from Simulated Ash Basin Water by CWTS.

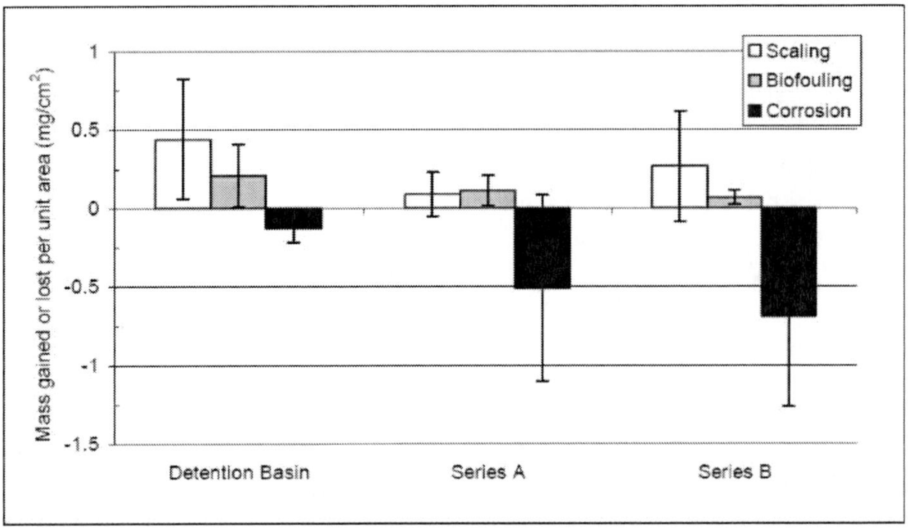

Figure 9-4. Scaling, Biofouling, and Corrosion Potential when Constructed Wetlands Treat Ash Basin Waters.

Other Issues

These two studies address water that would normally leave the fence line of the power plant. However, many NETL Existing Plants Program-funded water projects in research areas outside the non-traditional waters technology area consider other innovative methods for water recovery within the fence line, including:

- "A Synergistic Combination of Advanced Separation and Chemical Scale Inhibitor Technologies for Efficient Use of Impaired Water as Cooling Water in Coal-Based Power Plants," at Nalco Company.
- "Application of Pulsed Electrical Fields for Advanced Cooling and Water Recovery," at Drexel University.
- "Water Extraction from Coal-Fired Power Plant Flue Gas," by the University of North Dakota Energy and Environmental Research Center.
- "Recovery of Water from Boiler Flue Gas," by Lehigh University.
- "Use of Coal Drying to Reduce Water Consumed in Pulverized Coal Power Plants," by Lehigh University.
- "Transport Membrane Condenser for Water and Energy Recovery from Power Plant Flue Gas," by Gas Technology Institute.

In recent months, the lack of Federal regulations on the containment and disposal of coal ash pond waste has been a news item. EPA has been studying possible regulations of coal ash for 28 years but has not designated it as a hazardous waste. If future regulations demand that ash be stored dry in lined landfills, this source of non-traditional water might no longer be available.

CONCLUSION

Currently two studies are underway to examine the use of non-traditional waters that would normally be discharged from thermoelectric power plants. Both studies examine the use of ash pond waters. The Clemson University study has focused on using constructed wetlands treatment systems to treat the waters, and the University of Pittsburgh study has conducted bench-scale testing to determine appropriate chemical treatments.

Although ash pond waters are not discharged from the plant, researchers compared them to NPDES and WQC to represent their relative toxicities. The studies indicated that untreated ash pond waters typically exceed NPDES permit and WQC for arsenic, cadmium, chromium, copper, mercury, selenium, and zinc. Biofouling was not an issue of concern for power plant reuse, and bench-scale testing showed untreated waters were close to, or only slightly exceeded, the target criterion. If chlorine doses were kept low, the corrosion rates of the ash pond waters remained low. Scaling potential is also being studied.

The CWTS series at Clemson were found to decrease arsenic, chromium, mercury, selenium, and zinc concentrations below the detection limits. In addition, toxicity, scaling potential, and biofouling potential were all decreased in the treated waters compared to the untreated as basin waters. However, corrosion potential did increase. The University of

Pittsburgh study found that maintenance of chlorine at 0.5 ppm levels was optimal and that PMA, PTBC, and a combination (PMA:PBTC at 2:1) all afforded roughly the same scaling control. The use of PAA was not as effective.

Future Federal legislation regarding coal ash ponds should also be considered when examining the viability of recovering ash pond waters. Recent legislative efforts, as well as consideration of rules by the EPA, may affect future impoundment methods.

SECTION 10. LOCATING SOURCES OF NON-TRADITIONAL WATERS

Section 9 described an ongoing SNL project that is assessing power demand options in regions of water stress with regard to water produced from saline aquifers if CO2 is injected as part of a future carbon management program.[65] Earlier work examined aquifers in the Southwest, but efforts now are focusing on the Southeast and how specific evaluation metrics should be applied to characterize these water supplies. The study aims to consolidate the analysis framework into metrics, processes, and modeling aspects that can be applied across various regions of the United States. When that work is done, it will be important to relate its findings regarding nontraditional water availability to geographic locations for existing and planned thermoelectric power plants.

The proximity to non-traditional waters is an important issue when considering their use. Therefore, the following two GIS projects are underway that will help match thermoelectric power plants to nearby acceptable sources of non-traditional waters:

- "Optimization of Cooling Water Resources for Power Generation," (FT40320-01.24 and NT43291-05.1) conducted by the UND EERC that is developing a GIS-based web application decision support system for the Northern Great Plains Water Consortium.
- "Internet Based, GIS Catalog of Non-Traditional Sources of Cooling Water for Use at America's Coal-Fired Power Plants,"[66] (NT0005957) conducted by Arthur Langhus Layne, LLC (ALL Consulting) that will develop a nationwide catalog of the location, quantity, and quality of non-traditional water sources relative to power plants.

UND EERC is developing a decision support system[67] that uses web interfaces in conjunction with GIS tools to allow utilities to assess the availability of suitable water supplies in the Dakotas and Minnesota. These three states include a spatial extent that covers two EPA regions, different water rights doctrines (Section 3 of this document), and information management by both watershed districts and county water boards. The GIS-based information has been gathered from a number of sources, such as:

- Surface waters (USGS)
- Groundwater (individual states)
- Non-traditional waters (USGS and states)
- Water quantity (USGS and states)
- Water quality (EPA STORET database)

A sample screen shot displaying the data distribution is shown in Figure 10-1 and includes aquifers, rivers, and wastewater treatment plants. Both real-time and historical surface water flows can be tracked through the system using live links to USGS gauging station data (Figure 10-2). Clicking the mouse on a point provides the user access to detailed information about the sites through online links.

Although the geographic distribution of aquifer-bearing material is often known, the potential yields of the aquifers have not always been quantified. Figure 10-3 shows some aquifer yield information for various locations in the three-state area; the largest data gap has been groundwater yield information for South Dakota. The three states use different methods for calculating and reporting aquifer yield potential. However, quantitative information about surface non-traditional waters is often better known. For example, Figure 10-4 illustrates the municipal wastewater discharge rates at facilities discharging more than 1 million gallons per day. Researchers are also integrating the EPA's STORET database information to help determine surface water quality.

Figure 10-1. Data Distribution in the UND EERC Decision Support System.

In Phase 2 of the study, links to laws and tribal rights information will be included. Researchers plan to expand the geographic regions and include quantifiable data for produced waters (Section 6) and those from saline aquifers (Section 8). In addition to these GIS studies, Section 6 and Section 8 briefly discuss issues associated with access and collection of produced waters. Interested readers are encouraged to follow these discussions in order to understand the choices that were made about modes of transport, where to pre-treat waters along the pipeline, and how investigators evaluated the use of abandoned pipelines.

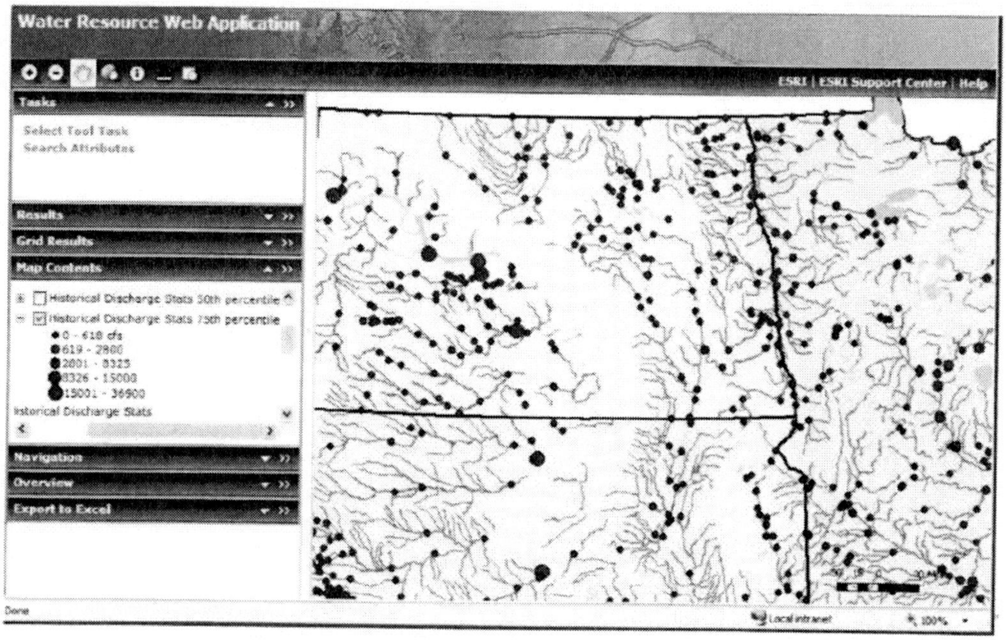

Figure 10-2. Historical Stream Flow Data Displayed by the UND EERC Decision Support System.

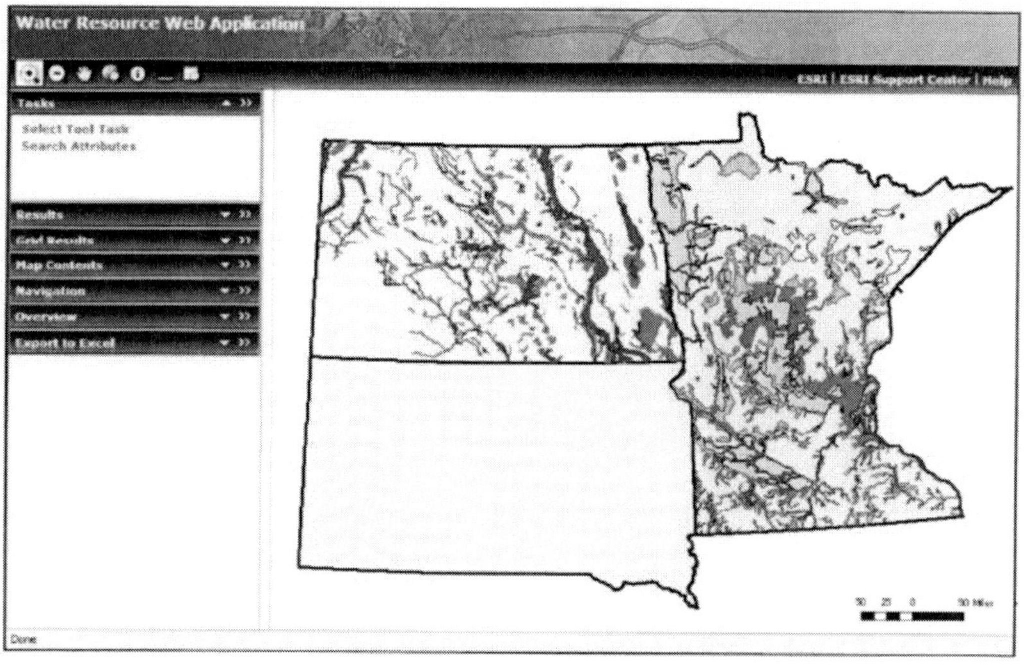

Figure 10-3. Aquifer Yield Information in the UND EERC Decision Support System (data gap for groundwater yield information for South Dakota).

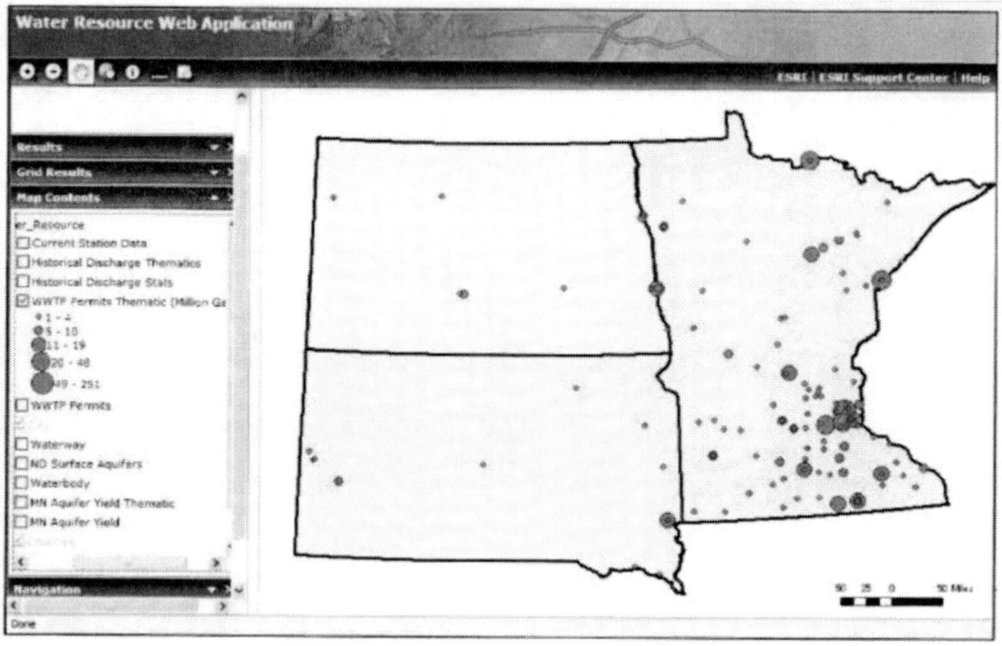

Figure 10-4. Municipal Wastewater Discharge Rates More Than 1 Million Gallons Per Day in the UND EERC Decision Support System.

ALL Consulting has begun an internet-based GIS effort to associate power plants with nontraditional water sources across the lower 48 states. The project advisory council includes members of the Ground Water Protection Council and the USGS. The Ground Water Protection Council is a national association of state ground water and underground injection control agencies whose mission is to promote protection and conservation of ground water resources for all beneficial uses.

In this project, ALL Consulting plans to:

- Identify the locations and water needs of coal-fired power plants in the lower 48 states.
- Identify the locations, quality, and volumes of non-traditional water sources.
- Create an internet-based GIS catalog of non-traditional sources of cooling water connecting the two geodatabases.

Information about power plant water usage will be collected by the Ground Water Protection Council. ALL Consulting is working with other NETL contractors and state officials to populate the geodatabase for the non-traditional sources of water. By clicking on a map location, users will be able to visualize the potential water sources, quality, volumes, and horizontal distance to those waters. The tool is intended to make users aware of water options and allow them to assess the costs for supplementing or replacing their current freshwater supplies.

APPENDIX A. DATABASE OF POWER PLANTS USING RECLAIMED WATER[2] (LISTED BY STATE)

Water User	Location	State	Volume	Water Source[2]	Location	Use of Water	Starting Date	Comments	Source
Arizona Public Service Palo Verde Nuclear Plant	Wintersburg	AZ	55 MGD	91st Avenue and Tolleson WWTPs	Phoenix and Tolleson, AZ	cooling tower makeup	1980	Tertiary treatment performed at power plant; all blowdown is evaporated in ponds.	Palo Verde Energy Information Center undated
Salt River Project – K-7 Plant	Tempe	AZ	3.1 MGD	Tempe Kyrene WRF	Tempe, AZ	cooling tower makeup			Salt River Project/City of Tempe 2005 (available at: www.azwater.gov/dwr/Content/Hot_Topics/files/SRP_K7-Tempe_KRF.pdf).
Arizona Public Service – Redhawk Power Station	Arlington	AZ	3.9 MGD	Tolleson WWTP	Tolleson, AZ	cooling tower makeup	2001	Water is sent via pipeline and is treated at the Palo Verde plant. It is then piped to Redhawk.	Tedder 2007
Burbank Power & Water – Olive Plant	Burbank	CA	0.1 MGD combined for Olive and Lake One plants	Burbank WWTP	Burbank, CA	cooling tower makeup	1967	The Olive and original Magnolia plants started using reclaimed water in 1967. The Lake One plant started later. The original Magnolia plant is now out of service, while the other two are on intermittent stand-by service.	Babayan 2007
Burbank Power & Water – original Magnolia Plant	Burbank	CA	not currently in service	Burbank WWTP	Burbank, CA	cooling tower makeup	1967	The Olive and original Magnolia plants started using reclaimed water in 1967. The Lake One plant started later. The original Magnolia plant is now out of service, while the other two are on intermittent stand-by service.	Babayan 2007
Burbank Water & Power – Lake One Plant	Burbank	CA	0.1 MGD combined for Olive and Lake One plants	Burbank WWTP	Burbank, CA	cooling tower makeup	2002	The Olive and original Magnolia plants started using reclaimed water in 1967. The Lake One plant started later. The original Magnolia plant is now out of service, while the other two are on intermittent stand-by service.	Babayan 2007
Southern California Public Power Authority – new Magnolia Plant	Burbank	CA	1.0 – 1.4 MGD	Burbank WWTP	Burbank, CA	cooling tower makeup	2005	The new Magnolia plant is at the same site as the original Magnolia Plant but is a modern plant. It is operated jointly by six cities.	Babayan 2007
Glendale Public Service – Grayson Plant	Glendale	CA	0.3 MGD	Glendale WWTP	Glendale, CA	cooling tower makeup	1979	Plant receives tertiary treated water.	CSWRCB 2000
Pacific Oroville Power – Cogen Facility	Oroville	CA	0.05 MGD	Oroville Region WWTP	Oroville, CA	cooling tower makeup	1989		CSWRB 2000; California Energy Commission undated (available at: http://www.energy.ca.gov/database/POWER_PLANTS.XLS).

[2] WWTP = wastewater treatment plant, WRF = water reclamation facility, and WPCF = water pollution control facility.

(Continued)

Water User	Location	State	Volume	Water Source[2]	Location	Use of Water	Starting Date	Comments	Source
Spadra Gas-to-Energy Plant	Pomona	CA	0.03 MGD	Pomona WRF	Pomona, CA	cooling tower makeup	1991	Plant burns gas from a closed landfill to generate power.	CSWRB 2000; California Energy Commission undated (available at: http://www.energy.ca.gov/database/POWER_PLANTS.XLS).
Puente Hills Energy Recovery Plant	Whittier	CA	0.5 MGD	San Jose Creek WRP	near Whittier, CA	cooling tower makeup	1994	Plant burns gas from a closed landfill to generate power.	CSWRB 2000, California Energy Commission undated (available at: http://www.energy.ca.gov/database/POWER_PLANTS.XLS).
Northern CA Power Agency – CT2 Project	Lodi	CA	0.08 MGD	White Slough WPCF	Lodi, CA	boiler feed	1996		CSWRB 2000; California Energy Commission undated (available at: http://www.energy.ca.gov/database/POWER_PLANTS.XLS).
Delta Energy Center	Pittsburg	CA	7.7 MGD – total for Delta and Los Medanos Energy Centers	Delta Diablo Sanitation District reclamation plant	Antioch, CA	cooling tower makeup	2001		Delta Diablo Sanitation District 2001 (available at: http://www.ddsd.org/pdfs/RWF_Brochure.pdf).
Los Medanos Energy Center	Pittsburg	CA	7.7 MGD – total for Delta and Los Medanos Energy Centers	Delta Diablo Sanitation District reclamation plant	Antioch, CA	cooling tower makeup	2001		Delta Diablo Sanitation District 2001 (available at: http://www.ddsd.org/pdfs/RWF_Brochure.pdf).
Geysers geothermal field – 19 Calpine geothermal power plants and several others operated by the Northern California Power Agency	Santa Rosa	CA	11 MGD	Santa Rosa WRF	Santa Rosa, CA	injected to maintain geothermal steam pressure	2003	This project will transport 11 million gallons of tertiary-treated reclaimed water from the City of Santa Rosa to the northwestern part of the Geysers steam field, where it will be injected in the geothermal reservoir.	GRDA and California Energy Commission undated (available at: http://www.energy.ca.gov/geothermal/fact_sheets/geothermal_projects/SANTA_ROSE_RECLAIMED_H20.PDF).
Geysers geothermal field – 19 Calpine geothermal power plants and several others operated by the Northern California Power Agency	Lake County	CA	8 MGD	Middletown, Southeast Regional, Northwest Regional, and Clearlake Oaks WWTPs	Lake County, CA	injected to maintain geothermal steam pressure	1997	This is a coordinated project to pipe reclaimed water from several WWTPs to a reservoir adjacent to the Geysers geothermal field. The water is injected to increase steam production.	Lake County California undated (available at: http://www.co.lake.ca.us/Government/DepartmentDirectory/Special_Districts/Wastewater_Systems/Effluent_Pipeline.htm).
Platte River Power Authority – Rawhide Energy Station	Wellington	CO	3.8 MGD	Drake WRF	Ft. Collins, CO	cooling	1984	The reclaimed water is piped to a 500-acre cooling pond at the Rawhide site; the cooling pond operates as a recirculating cooling system.	EPA 2004; fc.gov 2007 (available at: http://fcgov.com/wastewater/facilities.php); Platte River Power Authority 2007 (available at: http://prpa.org/energysources/rawhidehistory.htm).

Water User	Location	State	Volume	Water Source[a]	Location	Use of Water	Starting Date	Comments	Source
City of Lakeland – McIntosh Plant	Lakeland	FL	6 MGD	Lakeland WWTP	Lakeland, FL	cooling tower makeup	1983	Tertiary treatment is performed at power plant.	MPPRP 1997; Florida Department of Environmental Protection 2007a (available at: http://www.dep.state.fl.us/water/reuse/industry.htm).
Ogden Martin – Hillsborough County Solid Waste to Energy Recovery Facility	Tampa	FL	0.7 MGD	Hillsborough Co. Falkenburg WRF	Tampa, FL	cooling tower makeup	mid-1980s		EPA 2004; Duncan 2007
Southern Company – Oleander Power Project	Cocoa	FL	0.013 MGD	Jerry Sellers WRF	Cocoa, FL	cooling tower makeup			Harrell 2007
Vero Beach Municipal Power Plant	Vero Beach	FL	can use up to 15 MGD; recent maximum is 2 MGD	Vero Beach WWTP	Vero Beach, FL	cooling tower makeup	1992		Morris 2007
Ogden Martin – Lee County	Ft. Myers	FL	0.0 MGD	Ft. Myers City Central WWTP	Ft. Myers, FL	cooling tower makeup	1994		Curiel 2007
Progress Energy – Hines Energy Complex	Bartow	FL	1.8 MGD	Bartow WWTP	Bartow, FL	cooling ponds	1995		Adcock 2007
Wheelabrator – McKay Bay Waste-to-Energy Plant	Tampa	FL	0.6 MGD	Howard F. Curren WWTP	Tampa, FL	cooling tower makeup	1985		Varghes 2007
Wheelabrator – North Broward Waste-to-Energy Plant	Pompano Beach	FL	1.0 MGD	Broward County North Regional Plant	Pompano Beach, FL	cooling tower makeup	1991		Aliseo 2007
Tampa Electric Co. – Big Bend Plant	Tampa	FL	2 – 3.5 MGD	Hillsborough Co. South County Regional Facility and Falkenburg WRF	Tampa, FL	cooling tower makeup; air pollution control	1994		Duncan 2007
JEA – Northside Plant	Jacksonville	FL	1.9 MGD	JEA – District II WRF	Jacksonville, FL	cooling tower makeup			JEA undated (available at: http://www.jea.com/community/reclaim.asp).
Tallahassee – Purdom Generating Station	St. Marks	FL	0.75 MGD	St. Marks Powder	St. Marks, FL	cooling tower makeup	2002	This arrangement is unusual in that treated industrial wastewater from a gunpowder manufacturing facility is sent to the power plant for reuse.	King 2007

(Continued)

Water User	Location	State	Volume	Water Source[a]	Location	Use of Water	Starting Date	Comments	Source
Pasco County Resource Recovery Facility	Spring Hill	FL	0.2 MGD	Pasco County Master Reuse System	Pasco Co., FL	cooling tower makeup	mid-1990s	The county combines reclaimed water from several treatment plants into a combined county wide system.	Wright 2007
Orlando Utilities – C.H. Stanton Plant	Orlando	FL	10 – 11 MGD	Eastern WRF	Orlando, FL	cooling tower makeup	1997		EPA 2004; Castro 2007
Kissimmee Utility Authority – Cane Island Plant	Kissimmee	FL	1 – 2.3 MGD	South Bermuda WRF and Camelot WRF	Kissimmee, FL	cooling tower makeup	1990s	Two water reclamation plants combine reclaimed water into a long pipeline. Two separate power plants use portions of the water.	Burton 2007
Progress Energy – Intercession City Plant	Kissimmee	FL	0.03 – 0.2 MGD	South Bermuda WRF and Camelot WRF	Kissimmee, FL	air pollution control	1990s	Two water reclamation plants combine reclaimed water into a long pipeline. Two separate power plants use portions of the water	Burton 2007
Progress Energy Cogeneration Plant – Univ. of Florida	Gainesville	FL	0.4 MGD	Univ. of Florida WWTP	Gainesville, FL	cooling tower makeup	1994		UF Office of Sustainability 2008 (available at: http://www.sustainable.ufl.edu/operationdetails.asp?op=0); Florida Department of Environmental Protection 2007a (available at http://www.dep.state.fl.us/water/reuse/industry.htm).
Pinellas County Utilities – Waste-to-Energy Plant	St. Petersburg	FL	1.7 MGD	City of Largo WWTP	Largo, FL	cooling tower makeup	early 1990s		Black 2007
Alliant Energy	Clear Lake	IA	1.3 MGD	Clear Lake Sanitary District	Clear Lake, IA	cooling tower makeup	2003	Company paid for WWTP upgrades to tertiary quality. 60–80% of the water is evaporated and the blowdown is sent back to the WWTP.	WEF 2006
Sunflower Electric – Garden City Plant	Garden City	KS	0.5 MGD	Garden City	KS	cooling tower makeup	2007 (?)	They will use 90% treated sewage and 10% well water.	Carlson 2007
Wheelabrator – North Andover Waste to Energy Plant	North Andover	MA	0.5 0.66 MGD	Greater Lawrence Sanitation District	North Andover, MA	cooling tower makeup	1985		Marrinan 2007
Brayton Point	Somerset	MA	1.3 MGD	Somerset water pollution control plant	Somerset, MA	scrubber	2008 estimated		Metropolitan Area Planning Council 2005
Millennium Power Partners	Charlton	MA	0.4 MGD	Southbridge WWTP	Southbridge, MA	cooling tower makeup	~2004		Metropolitan Area Planning Council 2005; Grenier 2007

Water User	Location	State	Volume	Water Source[3]	Location	Use of Water	Starting Date	Comments	Source
Panda-Brandywine	Brandywine	MD	1.5 MGD	Mattawoman WWTP	Mason Springs, MD	cooling tower makeup	1997	The reclaimed water is sent through a 17-mile pipeline.	MPPRP 1997
Bethlehem Steel – Sparrows Pt.	Baltimore	MD	see comments	Back River WWTP	Baltimore, MD	process	early 1970s	Bethlehem Steel was sold under bankruptcy in 2003; the reclaimed water volume, at one time 100 MGD, is greatly decreased now because of steel mill downsizing	EPA 2004
Caledonia Operating Services – Caledonia Plant	Steens	MS	0.5 MGD	Columbus Light and Water WWTP	Columbus, MS		some time after 2002		Barksdale 2007; Cogentrix Energy, Inc., undated (available at: http://ww.willbros.com/_filelib/FileCabinet/Projects/North_America/Caledonia%20Water.pdf?FileName=C108.pdf).
AES – Granite Ridge Plant	Londonderry	NH	4 MGD	Manchester WWTP	Manchester, NH	cooling tower makeup	2002		NH Society of Professional Engineers 2002
PSEG – Linden Station	Linden	NJ	12 MGD	Linden/Roselle Sewage Authority	Linden, NJ	cooling tower makeup	2006	Plant provides additional treatment before using. Any unused water is returned to the WWTP.	EIA 2005; *Power Magazine* 2006
PSEG – Bergen Station	Ridgefield	NJ	0.6 MGD	Bergen County Utilities Authority WWTP	Little Ferry, NJ	cooling tower makeup	2002		Bergen County Utilities Authority 2006 (available at: http://www.bcua.org/WPC_VT_WasteWaterReUse.htm).
PNM – Luna Energy Facility	Deming	NM	1 MGD	Deming WWTP	Deming, NM	cooling tower makeup	2007	Plant commissioned in 4/06 but filtration system failed. The company's plan was to recommission in the spring of 2007.	Lavery 2007
Nevada Power Company – Clark Station	Henderson	NV	varies monthly – 0.15 to 2.7 MGD	Clark County Water Reclamation District	Las Vegas, NV	cooling tower makeup		The power company provides additional treatment before use.	Morgan 2007
Nevada Power Company – Sunrise Station	Las Vegas	NV	0.09 – 0.3 MGD	Las Vegas WPCF	Las Vegas, NV	cooling tower makeup	early 1990s		Miller 2007
Lancaster County Resource Recovery Facility	Bainbridge	PA	0.8 MGD	Elizabethtown WWTP	Elizabeth-town, PA	cooling tower makeup	1991	The facility has holding ponds with a 2.5 million gallon capacity.	LCSWMA undated (available at: http://www.lcswma.org/resourceRecovery.asp); EPA 2004.
American National Power – Hays Energy Project	San Marcos	TX	0.3 MGD	San Marcos	San Marcos, TX	cooling tower makeup	2002		EPA 2004; Longoria et al. 2000

(Continued)

Water User	Location	State	Volume	Water Source[2]	Location	Use of Water	Starting Date	Comments	Source
Garland Power & Light – Spencer Road Power Plant	Denton	TX	1.0 MGD	Pecan Creek WRF	Denton, TX	cooling tower makeup	mid-1970s	Facility may have been shut down in 2002	City of Denton, Texas 2003 (available at: http://www.cityofdenton.com/pages/mypovewhitepaperssolidwwaterwater.cfm?object=3481&folderID=258&action=files).
Xcel Energy – Harrington Plant	Amarillo	TX	15 MGD split between Nichols and Harring-ton	Amarillo WWTP	Amarillo, TX	cooling tower makeup			Xcel Energy undated *a* (available at: http://www.xcelenergy.com/XLWEB/CDA/0,3080,1-1-1_1875_4797_4063-3613-0_0_0-0,00.html).
Xcel Energy – Jones Plant	Lubbock	TX	3 – 5 MGD	Lubbock WWTP	Lubbock, TX	cooling tower makeup			Xcel Energy undated *b* (available at: http://www.xcelenergy.com/XLWEB/CDA/0,3080,1-1-1_1875_4797_4063-3614-0_0_0-0,00.html).
Xcel Energy – Nichols Plant	Amarillo	TX	15 MGD split between Nichols and Harring-ton	Amarillo WWTP	Amarillo, TX	cooling tower makeup			Xcel Energy undated *c* (available at: http://www.xcelenergy.com/XLWEB/CDA/0,3080,1-1-1_1875_4797_4063-3615-0_0_0-0,00.html).
El Paso Electric – Newman Plant	El Paso	TX	2.4 MGD	El Paso Water Utilities - Fred Hervey WWTP	El Paso, TX	cooling tower makeup	1991		Mauricio 2007; El Paso Water Utilities 2007 (available at: http://www.epwu.org/wastewater/fred_hervey_reclaimation.html#).
Suez Energy Resources – Ennis-Tractebel Plant	Ennis	TX	1 – 1.5 MGD	Oak Grove WWTP	Ennis, TX	cooling tower makeup	2001		Hester 2007
GE Energy – Fox Energy Center	Kaukauna	WI	0.8 MGD	Heart of the Valley Metropolitan Sewerage District	Kaukauna, WI	cooling tower makeup	2005	Use is intermittent – this is a peaking plant with most use in the summer.	Guerts 2007, Johnson 2007

End Notes for Section 1

[1] U.S. Census Bureau, Table T1. Population Estimates, factfinder.census.gov, last accessed March 23, 2009.

[2] Kenny, J.F., N.L. Barber, S.S Hutson, K.S. Linsey, J.K. Lovelace, and M.A. Maupin, Estimated Use of Water in the United States in 2005, U.S. Geological Survey Circular 1344, 2009, http://pubs.usgs.gov/circ/1344/pdf/c1344.pdf/

[3] Solley, W.B., R. R. Pierce, and H.A. Perlman, Estimated Use of Water in the United States in 1995, U.S. Geological Survey Circular 1200, 1998, http://water.usgs.gov/watuse/pdf1995/html/, last accessed March 25, 2009.

[4] B. Carney, T. Feeley, and A. McNemar. "Department of Energy, National Energy Technology Laboratory, Power Plant-Water R&D Program." Presented at the Water-Energy Nexus Conference in December 2008. Available at: http://www.netl.doe.gov/technologies/coalpower/ewr/water/pdfs/NETL%20Paper%20Unesco%20Conference.pdf.

End Notes for Section 3

[5] U.S. Environment Protection Agency and U.S. Agency for International Development, Guidelines for Water Reuse, EPA/625/R-04/108, September 2004.

[6] University of Pittsburgh. *Reuse of Treated Internal or External Wastewaters in the Cooling Systems of Coal-Based Thermoelectric Power Plants*; Quarterly Report for DOE/NETL Project No. DE-FC26-06NT42722; December 2006.

[7] Office of the Secretary of State Division of State Documents. Title 26: Department of the Environment. http://www.dsd.state.md.us/comar/subtitle_Sections/26_Sections.htm, last accessed March 11, 2009.

[8] Pennsylvania Department of Environmental Protection, Reuse of Treated Wastewater Guidance Manual, http://www.dep.state.pa.us/TechnicalGuidance/Draft_technical_guidance.asp, last accessed March 11, 2009.

[9] Department of Ecology, State of Washington. "Preliminary Baseline Rule Language Comments." http://www.ecy.wa.gov/programs/wq/reclaim/comments/COMMENTS.html, last accessed March 11, 2009.

[10] Wyoming Secretary of State, Standards for the Reuse of Treated Wastewaters Section 21. http://soswy.state.wy.us/Rules/RULES/2804.pdf, last accessed March 11, 2009.

[11] J.L. Gillette and J.A. Veil. Identification of Incentive Options to Encourage the Use of Produced Water, Coal Bed Methane Water, and Mine Pool Water. September 2004.

End Notes for Section 4

[12] Both freshwater (approximately 70 percent of U.S. total) and saline water (approximately 30 percent of U.S. total) are currently used for thermoelectric generation. Saline water is used primarily at plants located along the coast that are equipped with once-through cooling systems.

[13] Kenny, J.F., N.L. Barber, S.S Hutson, K.S. Linsey, J.K. Lovelace, and M.A. Maupin, Estimated Use of Water in the United States in 2005, U.S. Geological Survey Circular 1344, 2009.

[14] USGS. *Estimated Use of Water in the United States in 1995*; USGS Circular 1200; 1998.

[15] DOE/NETL. *Estimating Freshwater Needs to Meet Future Thermoelectric Generation Requirements;* September 2008.

[16] A relatively small percentage of thermoelectric power plants in the United States utilize dry cooling systems that require a minimal supply of makeup water.

[17] Compared to the cooling water and FGD systems, there is a minimal quantity of makeup water required for the boiler, ash handling, wastewater treatment, and general plant wash down systems. As a result, there is no further discussion of these water uses in this Section.

[18] Parsons. *Power Plant Water Usage and Loss Study*; Report prepared for U.S. DOE/NETL; May 2007.

[19] Daniels, David. *Untangling the Complexities of Cooling Water Chemistry*; Power Magazine; September 2002.

[20] Nalco Company. *A Synergistic Combination of Advanced Separation and Chemical Scale Inhibitor Technologies for Efficient Use of Impaired Water as Cooling Water in Coal-Based Power Plants*; Quarterly Report for DOE/NETL Project No. DE-FC26-06NT42721; July 2006.
[21] Argonne National Laboratory. *Use of Reclaimed Water for Power Plant Cooling*; August 2007.
[22] Mortensen, Ken. *How to Manage Cooling Tower Water Quality*; RSES Journal; May 2003.
[23] SPX Cooling Technologies. *Cooling Tower Fundamentals*; Published by Marley Cooling Technologies, Inc.; Second edition; Copyright 1998.
[24] Buecker, Brad. *Choosing the Right Cooling Tower Chemicals*; Power Engineering; July 1998.
[25] Nalco Company. *A Synergistic Combination of Advanced Separation and Chemical Scale Inhibitor Technologies for Efficient Use of Impaired Water as Cooling Water in Coal-Based Power Plants*; Quarterly Report for DOE/NETL Project No. DE-FC26-06NT42721; October 2006.
[26] EPRI. *Use of Degraded Water Sources as Cooling Water in Power Plants*; Report No. P500-03-110 prepared for California Energy Commission; October 2003.
[27] Benson, Lew. Carmeuse North America; Personal communication on September 10, 2007.
[28] Buecker, Brad. Kansas City Power & Light Company; Personal communication on September 10, 2007.
[29] EPRI. Use of Produced Water in Recirculating Cooling Systems at Power Generating Facilities; Semi-Annual Technical Progress Report for DOE/NETL Project No. De-FC26-03NT41906; October 2004. Available at: http://www.netl.doe.gov/technologies/coalpower/ewr/water/pp-mgmt/pubs/41906/41906Treatment%26DisposalAnalysis.pdf.
[30] Radian International. Electric Utility Engineer's FGD Manual, Volume I – FGD Process Design. Final Report for DOE Project No. DE-FG22-95PC94256; March 1996.
[31] Relative saturation is the ratio of the actual concentration to the theoretical saturation concentration based on the solution's chemical composition, pH, and temperature.

End Notes for Section 5

[32] EPA (U.S. Environmental Protection Agency), 2004. Guidelines for Water Reuse, EPA/625/R-04/108, U.S. Environmental Protection Agency and U.S. Agency for International Development. Available at : http://www.epa.gov/nrmrl/pubs/625r04108/625r04108.pdf.
[33] J.A. Veil. Use of Reclaimed Water for Power Plant Cooling, ANL/EVS/R-07/3, August 2007.
34 R.D. Vidic and D. Dzombak. Reuse of Treated Internal or External Wastewaters in the Cooling Systems of Coal-based Thermoelectric Power Plants: Final Technical Report September 2009. http://www.netl.doe.gov/technologies/coalpower/ewr/water/pp-mgmt/pubs/06550/42722FSRFG063009.pdf
[35] D. Dzombak R.D. Vidic and A.E. Landis. Use of Treated Municipal Wastewaters as Power Plant Cooling System Makeup Water: Tertiary Treatment versus Expanded Chemical Regimen for Recirculating Water Quality Management. Proposal submitted April 21, 2008.

End Notes for Section 6

[36] J.L. Gillette and J.A. Veil. Identification of Incentive Options to Encourage the Use of Produced Water, Coal Bed Methane Water, and Mine Pool Water. September 2004.
[37] Zammit, Kent. Semi-Annual Technical Progress Report October 1, 2005 to March 31, 2006: Use of Produced Water in Recirculating Cooling Systems at Power Generating Facilities. Deliverable Number 11 for DOE Award No. 41906.
[38] J.H. Rodgers *et al.* An Innovative System for the Efficient and Effective Treatment of Non-traditional Waters for Reuse in Thermoelectric Power Generation. Fifth Quarterly Progress Report for DOE Award DE-FG26-05NT42535. January 2007.
[39] Notice of Financial Assistance Award. Project Title: Reuse of Produced Water from CO2 Enhanced Oil Recovery, Coal-Bed Methane, and Mine Pool Waster by Coal-Based Power Plants. (NT0005343). September 2008.
[40] Sandia National Laboratories. Work statement on Nanofiltration Treatment Options project.
[41] University of Illinois. Statement of Project Objectives. DE-NT0005343.

End Notes for Section 7

[42] J.A. Veil, J.M. Kupar, and M.G. Puder. Use of Mine Pool Water for Power Plant Cooling. September 2003.

[43] West Virginia University. Strategies for Cooling Electric Generating Facilities Utilizing Mine Water: Technical and Economic Feasibility.

[44] C.A. Cravotta, K..B.C. Brady, A.W. Rose, and J.B. Douds. Frequency Distribution of the pH of Coal Mine Drainage in Pennsylvania. In D.W. Morganwalp and Buxton, eds. U.S. Geological Survey Toxic Substances Hydrology Program – Proceedings of the Technical Meeting, Charleston, South Carolina, March 8-12, 1999. U.S. Geological Survey Water Resources Investigations Report 99-4018A.

[45] R.D. Vidic and D. Dzombak. Reuse of Treated Internal or External Wastewaters in the Cooling Systems of Coal-based Thermoelectric Power Plants. April 1, 2008 – June 30, 2008.

[46] R.D. Vidic and D. Dzombak. Reuse of Treated Internal or External Wastewaters in the Cooling Systems of Coal-based Thermoelectric Power Plants: Final Technical Report September 2009. http://www.netl.doe.gov /technologies/coalpower/ewr/water/pp-mgmt/pubs/06550/42722FSRFG063009.pdf.

[47] P. Ziemkiewicz, R. Herd, and B. Leavitt. Development and Demonstration of a Modeling Framework for Assessing the Efficacy of Using Mine Water for Thermoelectric Power Generation. January 2007.

[48] J.L. Gillette and J.A. Veil. Identification of Incentive Options to Encourage the Use of Produced Water, Coal Bed Methane Water, and Mine Pool Water. September 2004.

[49] G.R. Watzlaf and T.E. Ackman. Underground Mine Water for Heating and Cooling Using Geothermal Heat Pump Systems. Technical Communication, 2006.

End Notes for Section 8

[50] Kobos *et al.*, Sandia Report SAND2008-4037, Study of the Use of Saline Aquifers for Combined Thermoelectric Power Plant Water Needs and Carbon Sequestration at a Regional Scale: Phase I Report (June 2008).

[51] Kansas Geological Survey, NATCARB Homepage, http://www.natcarb.org/index_nc.html, last accessed March 14, 2009.

[52] Zammit, Kent. Semi-Annual Technical Progress Report October 1, 2005 to March 31, 2006: Use of Produced Water in Recirculating Cooling Systems at Power Generating Facilities. Deliverable Number 11 for DOE Award No. 41906.

[53] National Energy Technology Laboratory, "NETL-Existing Plants Water Projects Meeting Presentations," stored at http://www.netl.doe.gov/technologies/coalpower/ewr/water/pdfs/NETL%20Existing%20Plants% 20Water%20Projec ts%20Meeting%20Presentations.pdf, last accessed March 13, 2009.

[54] GE Water and Process Technologies. *Electrodialysis Reversal (EDR)*, http://www.gewater.com/products /equipment/ed_edr_edi/edr.jsp, last accessed March 17, 2009.

[55] D. Whisenhunt, J. Melzer, and A. Shetty, "Minimization of Fresh Water Usage at Coal-Fired Power Plants." Presented at October 2008 meeting.

[56] Department of Energy/National Energy Technology Laboratory (DOE/NETL). *Estimating Freshwater Needs to Meet Future Thermoelectric Generation Requirements: 2007 Update Forecasts.* DOE/NETL-400/2007/1304. http://www.fypower.org/pdf/DOE_WaterNeedsAnalysis_2007.pdf.

[57] Department of Energy/National Energy Technology Laboratory (DOE/NETL). *Estimating Freshwater Needs to Meet Future Thermoelectric Generation Requirements: 2008 Update.* DOE/NETL-400/2008/1339. http://www.netl.doe.gov/technologies/coalpower/ewr/pubs/2008_Water_Needs_Analysis-Final_10-2-2008.pdf.

[58] Federal Register, Vol. 73, No. 144, 43492-43541, July 25, 2008. Regulatory development can be tracked at: http://www.epa.gov/safewater/uic/wells sequestration.html.

End Notes for Section 9

[59] Allyson Groff and Blake Androff, "Rahall Introduces Legislation to Regulate Coal Ash Disposal." January 14, 2009. http://resourcescommittee.house.gov/index.php?option=com_content&task=view&id=439&Itemid=27.

[60] J.H. Rodgers *et al.* An Innovative System for the Efficient and Effective Treatment of Non-traditional Waters for Reuse in Thermoelectric Power Generation. Fifth Quarterly Progress Report for DOE Award DE-FG26-05NT42535. January 2007.

[61] R.D. Vidic and D. Dzombak Reuse of Treated Internal or External Wastewaters in the Cooling Systems of Coal-Based Thermoelectric Power Plants. July 1, 2007-September 30, 2007, Quarterly Progress Report for DOE Award DE-FC26-06NT42722.

[62] R.D. Vidic and D. Dzombak Reuse of Treated Internal or External Wastewaters in the Cooling Systems of Coal-Based Thermoelectric Power Plants. January 1, 2008-March 31, 2008, Quarterly Progress Report for DOE Award DE-FC26-06NT42722.

[63] R.D. Vidic and D. Dzombak. Reuse of Treated Internal or External Wastewaters in the Cooling Systems of Coal-based Thermoelectric Power Plants: Final Technical Report September 2009. http://www.netl.doe.gov/technologies/coalpower/ewr/water/pp-mgmt/pubs/06550/42722FSRFG063009.pdf

[64] J.H. Rodgers *et al.* An Innovative System for the Efficient and Effective Treatment of Non-traditional Waters for Reuse in Thermoelectric Power Generation. Final Technical Report for DOE Award DE-FG26-05NT42535. November 28, 2008.

End Notes for Section 10

[65] National Energy Technology Laboratory, *Power Plant Water Management: Thermoelectric Power Plant Water Demands Using Alternative Water Supplies: Power Demand Options in Regions of Water Stress and Future Carbon Management.* http://204.154.137.14/technologies/coalpower/ewr/water/pp-mgmt/regional.html, last accessed March 11, 2009.

[66] National Energy Technology Laboratory, *Power Plant Water Management: Internet-Based, GIS Catalog of Non-Traditional Sources of Cooling Water for Use at Coal-Fired Power Plants.* http://www.netl.doe.gov /technologies/coalpower/ewr/water/pp-mgmt/gis.html, last accessed March 13, 2009.

[67] National Energy Technology Laboratory, "NETL-Existing Plants Water Projects Meeting Presentations," stored at http://www.netl.doe.gov/technologies/coalpower/ewr/water/pdfs/NETL%20Existing%20Plants %20Water%20Projec ts%20Meeting%20Presentations.pdf, last accessed March 13, 2009.

In: Alternative Water Use for Power Plant Applications
Editor: Eric N. Grady
ISBN: 978-1-62808-986-8

Chapter 2

THERMOELECTRIC POWER PLANT WATER DEMANDS USING ALTERNATIVE WATER SUPPLIES: POWER DEMAND OPTIONS IN REGIONS OF WATER STRESS AND FUTURE CARBON MANAGEMENT*

Peter H. Kobos, Jim L. Krumhansl, Thomas A. Dewers, Geoffrey T. Klise, Brian P. Dwyer, Vincent C. Tidwell, Richard Kottenstette, David J. Borns and Andrea McNemar

ABSTRACT

Throughout the United States, many sizable saline formations may have the potential to hold carbon dioxide (CO_2), a capability that could be used to manage CO_2 emissions from power plants. This analysis examined the southeastern U.S. to determine where saline formations exist that could be used both to hold CO_2, as well as to provide saline waters for treatment to develop a coupled-use system to extract the water, treat it, and then use it to help with power plant cooling efforts. While this is a high-level study, the results indicate that limited saline formation/well data restricted the ability to adequately assess many regions of the southeastern U.S. In Texas, however, the Carrizo-Wilcox system met the size, location, depth, and other criteria to potentially host a coupled CO_2 sequestration system along with water extraction and treatment for use. Using the analytical framework applied in this analysis, other regions of the country and site-specific test cases may benefit from the insight provided by a coupled-use (or similar) system to increase the value of CO_2 storage to meet growing regional water stress on the surface.

---

* This is an edited, reformatted and augmented version of National Energy Technology Laboratory, Publication No. DOE/NETL-FWP 08-014053; SAND2011-5808P, dated August 2010.

ACRONYMS AND ABBREVIATIONS

$	dollars (United States)
°C	degrees Celsius
BWRO	brackish water reverse osmosis
CBM	coal bed methane
CCS	carbon capture and storage
CO2	carbon dioxide
DOE	Department of Energy
eGRID	Emissions & Generation Resource Integrated Database
EPA	Environmental Protection Agency
Fm	formation
ft	feet
ft^3	cubic feet
g	grams
GAM	Groundwater Availability Model
g/cm^3	grams per cubic centimeter
gpm	gallons per minute
H_2O	water
GW	Groundwater
HEROTM	high-efficiency reverse osmosis
IGCC	integrated gasification combined cycle
JTU	Jackson Turbidity Unit
kg/s	kilograms per second
km	kilometers
kWh	kilowatt-hours
m	meters
m^2	square meters
m^3	cubic meters
Mg	magnesium
MGD	millions of gallons per day
MGY	millions of gallons per year
mi	miles
MPa	megapascals
MW	megawatts
Na	sodium
NATCARB	Distributed National Carbon Sequestration Database and Geographical Information System
NETL	National Energy Technology Laboratory
O&M	Operational and maintenance
P	pressure
Pa	pascals
PC	pulverized coal
ppm	parts per million
RO	Reverse osmosis

s	seconds
SECARB	Southeast Regional Carbon Sequestration Partnership
SNL	Sandia National Laboratories
SDWA	Safe Drinking Water Act
s.u.	Standard Units
TCF	trillion cubic feet
TDS	total dissolved solids
TWDB	Texas Water Development Board
µS/cm	microSiemens per centimeter
U.S.	United States
USBR	United States Bureau of Reclamation
WECS	Water, Energy, and Carbon Sequestration
yr	years

1. INTRODUCTION

Expanding power demands across the country may face increasingly restrictive water supplies and environmental considerations in certain regions, including carbon management in the coming decades. Capitalizing on previously untapped sources of water, possibly unsuitable for potable demands may relieve growing concerns with water, environmental, and other constraints on power demand growth. Produced waters, nonpotable groundwater, and previously underutilized waters provided by local municipalities (already being used in some parts of the country) may prove an increasingly valuable resource. Regions of the country, especially in the west and southwestern United States, will likely see an increase in demand for lower-quality waters in an effort to save higher-quality waters for more valuable uses such as meeting agriculture, commercial, and residential demand. At the same time, other regions, such as the southeastern United States, may see increasing pressure to supply power for a growing regional economy.

A better regional understanding is needed of power plant water needs, suitability for existing and new water treatment and power plant cooling technologies, and regional power demand growth with an eye on a possible future with carbon dioxide constraints. With this better understanding, several technological, policy, and water resource management questions can be addressed when planning for sustained (or growing) power plant water demands. Of paramount importance within this project is to identify where potential value may be added to meet water stress issues and enhance the sustainability of economical power supplies. Figure 1-1 illustrates the overall concept of a coupled CO_2 sequestration system, with an extracted and treated water system as described in Kobos et al. (2009).

Several water treatment and desalination issues must be addressed to assess the scale of potential water supplies, their relative cost (e.g., if they require treatment for an appropriate use), and regional considerations. One is to address the scale of nontraditional water supply availability in different regions of the country relative to the power plant demands. Coal, nuclear, and other large-scale power plants require sustained, low-cost supplies of water for cooling. There may be, however, an equally high demand for additional water supplies from competing uses. Therefore, there may be a trend to site new (and replacement) power plants near sustainable water supplies that have less demand upon them such as existing, produced, or treated municipal water supplies. The economics of using these water resources for thermoelectric cooling potential in regions of the country with sizable supplies of otherwise underutilized waters should be assessed.

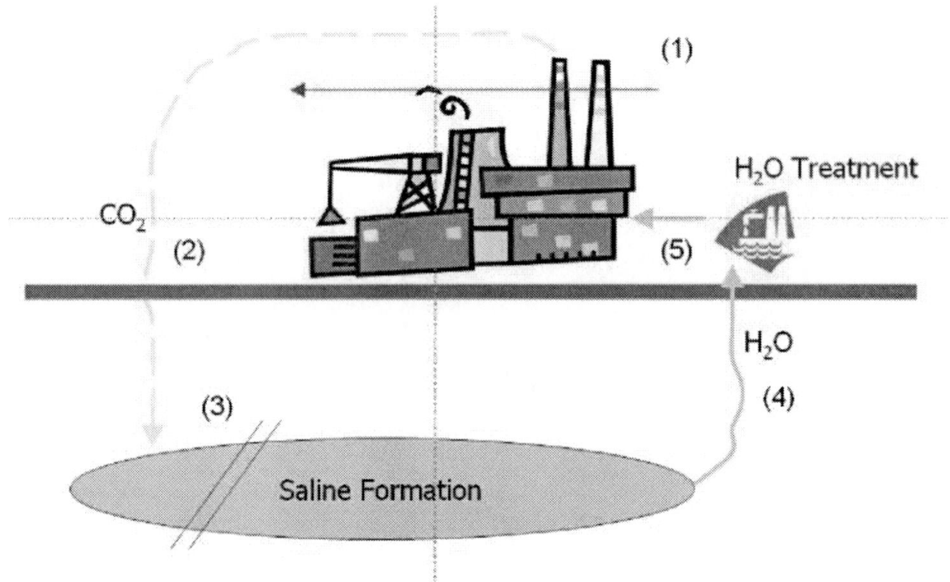

(1) Power Plant Metrics, (2) Carbon Sequestration System, (3) Saline Formation Geochemical and Geomodeling Assessment, (4) Water Extraction Analysis, (5) Water Treatment Analysis for Power Plant Cooling.

Figure 1-1. Coupled Carbon Dioxide Sequestration and Water Extraction and Treatment Conceptual Model Framework.

Additionally, several issues must be addressed regarding current and future power demand throughout the country. While there may be a regulatory drive to expand power supplies near existing power plants, some sites may be more suitable than others because of other factors. Of paramount importance will be the sustainability of water supplies, with consideration given to the full system's environmental issues. Water from geological saline formations, for example, will likely require some treatment prior to using it to cool power plants, but at an economic premium. Produced waters from fossil fuel production (e.g., coal-bed methane, mine pool waters, etc.) represent another source of potential cooling water. These sources, however, may prove challenging to fully characterize with respect to their overall size, uniformity of quality, potential environmental impacts, and thus long-term suitability as a sustainable resource for cooling water. Municipal waters previously underutilized, however, may also prove to be a unique source of cooling water.

This project systematically addresses each of these factors to determine (1) the applicability of the assessment framework to answer whether nontraditional water sources can be utilized for power plant cooling at the relative size and time scales necessary to provide an economical solution to constrained water supplies and (2) the identification of where market- and science-based solutions may offer relief to power plant demand that may be constrained by traditional water availability in certain regions of the U.S. by using water from geological saline formations. This will get to the very heart of the Energy-Water Nexus: how can power demand continue at current or increased levels while accounting for increasing water stress across many regions of the country?

The analysis addresses several central issues:

1. Down-select to areas within the southeastern U.S. to dovetail with potential pilot projects for carbon sequestration within National Energy Technology Laboratory's (NETL's) Southeast Regional Carbon Sequestration Partnership (SECARB) and additional regions of interest that are heavily reliant on coal-fired power plants to meet current and future demand loads;
2. Identify power plants in areas of interest to develop an analytical framework (e.g., a type of roadmap) to determine power demand solutions while accounting for potential increasing costs resulting from water demand stress for cooling options and potential carbon management issues;
3. Identify and characterize water treatment options for the regional "story" of the potential water supplies while accounting for decreasing water quality (and hence increasing treatment cost) options in the face of sustained or growing demand for power plant cooling;
4. Consolidate the framework into a package of metrics, processes, and modeling aspects that may be applied to other regions of the United States to first identify the regional, and then more site-specific, aspects of water-for-energy planning.

2. Geoassessment of Saline Formations within the Southeastern United States

2.1. Geoassessment Introduction

The area of the southeastern United States considered within SECARB has recently experienced rapid growth that stressed regional resources, including assuring adequate supplies of electrical power and potable water. Increasingly, surface water resources are allocated to community water needs. Economic growth often comes with an increased demand for electricity that, in turn, may require additional water for power plants under specific growth scenarios (DOE, 2006a, b, c).

Sandia National Laboratories (SNL) and NETL have undertaken a program that addresses both carbon dioxide sequestration and water supply issues. The working premise is that capturing CO_2 from power plant stack gases may become necessary, and storing CO_2 underground may be one of many carbon management options. The analysis also accounts for the fact that injecting CO_2 into a porous underground formation will likely displace saline waters in some formations that could become a valuable resource with the application of specific water treatment technologies.

The Environmental Protection Agency (EPA) is proposing new federal requirements under the Safe Drinking Water Act (SDWA) to address underground injection of CO_2 (EPA, 2008). Under these requirements, saline formations containing waters with less than 10,000 parts per million (ppm) are not to be considered for projects involving geologic CO_2

sequestration. The analysis does not consider further study on saline formations under this proposed limit, although mildly salty "brackish" waters in geological formations may be a likely source of water for the utility industry. From a water desalination and economic standpoint, the general range of saline waters under consideration range from 10,000 ppm to 35,000 ppm. Additionally, the economics of transportation and injection of CO_2 dictate that formations receiving the gas must be at least 2,500 feet (800 meters) below the surface (Allis et al., 2003, NETL, 2008a).

The first activity of the overarching analysis is to identify suitable geological formations deep enough to store CO_2 and contain saline waters for potential extraction, treatment, and use. Looking to the southeastern United States, the analysis begins by examining the area overseen by SECARB, one of NETL's partnership regions. This portion of the study reviewed these formations and determined whether potential coupled-use sites existed.

2.1.1. Geoassessment Introductory Summary

One of several aspects of the regional partnerships on carbon sequestration is to develop pilot tests of CO_2 injection processes. Thus, it was reasonable to ask if the projected pilot test sites within the SECARB region might also be amenable to analysis from the standpoint of coupled-use applications. The initial assessment found that only one of the four sites (the Black Warrior Basin) looks promising for consideration as a coupled-use site.

A second approach towards the SECARB region was to survey the whole designated area based on data gleaned from various government publications. A compilation of potential CO_2 sequestration resources provided by the Texas Bureau of Economic Geology (Hovorka et al., 2000) was particularly helpful in this endeavor. Areas in south-central Florida (centered on Polk County; Sprinkle, 1989[1]; USGS, 1999[2]) and south-central Texas (inland ~100 miles from the Gulf coast; Texas Water Development Board, February, 2003) were identified as likely to meet the relevant water quality and depth criteria. A third promising area in southeast Georgia (Sprinkle, 1989[1]; USGS, 1999[2]) was also identified, but water quality data from this region are virtually nonexistent, which restricted further analysis.

The Black Warrior Basin, the Texas site, and the Florida site have storage capacities of $2.6x10^{11}$ kg of CO_2 (Pashin et al., 2003), $4.3x10^{11}$ m^3 and $2.6x10^{14}$ kg @ 100% CO_2 Occupancy(assuming that the CO_2 density is 0.6 g/cm^3), and $1.04x10^{12}$ m^3 and $6.2x10^{14}$ kg @ 100% CO_2 Occupancy (assuming that the CO_2 density is 0.6 g/cm^3), respectively. Sequestration at the Texas and Florida sites would involve the "standard" process of injecting CO_2 into a porous, brackish, water-filled sandstone. Hence, the initial storage capacity is simply based on a geometric estimate of the amounts of rock (and enclosed porosity) that meet coupled-use criteria. In the Black Warrior Basin, CO_2 is injected in conjunction with the recovery of coal bed methane. Since the methane is released as CO_2 and is absorbed (by coal), the sequestration mechanism is different from when pore waters are mechanically displaced by the CO_2. The Black Warrior Basin estimate was obtained from a study (Pashin et al., 2003) that considers such processes.

The initial findings suggest that only a small fraction of the total SECARB designated area is amenable to a coupled-use approach in the strictest sense of the concept. The study also revealed, however, that the potential coupled-use resource base would increase dramatically if a shallow (around 1,500 feet deep) formation provided the brackish water while the CO_2 was injected into a deeper (e.g., below 2,500 feet to keep the CO_2 in a supercritical state), high-salinity geological formation to extract saline waters. Technical questions still requiring resolution include the following:

1. Will CO_2 injected into a layered sequence of gently dipping rocks eventually migrate updip and reappear at the surface?
2. What fraction of the porosity of a brine-filled sandstone aquifer will the CO_2 actually displace?
3. What parameters are appropriate for evaluating the CO_2 sequestration capacity at coal bed methane sites, as compared to brine-filled sandstones. Modeling the impact of injecting CO_2 on local groundwater quality at these sites is an additional topic deserving of study.

2.2. Details of Regional Geoassessment Analysis

The SECARB area comprises those states in the southeastern quadrant of the U.S. (Figure 2-1: lighter area—Alabama, Arkansas, Florida, Georgia, Louisiana, Mississippi, North Carolina, South Carolina, Tennessee, and Virginia plus counties in Kentucky, Texas, and West Virginia). This is a much larger area than the San Juan Basin (see Kobos et al., 2009) and contains a wider variety of geologic provinces to be evaluated. Along the western margin of the Appalachian Mountains are several basins that survived the mountain building and are filled with sediments (coal beds inter-layered with limestone, shale, and sandstone) and that are attractive CO2 sequestration targets.

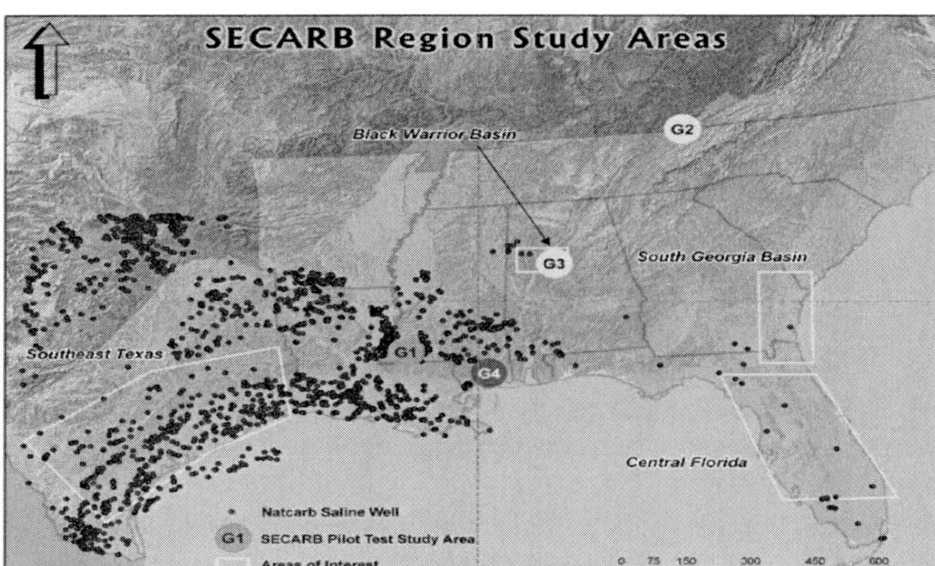

Data Source: NATCARB, 2008; NETL, 2008a, 2008b. Note: Regional designation with four pilot test site locations (NETL, 2008b). Cranfield (G1, green), Central Appalachian Basin Coal Test (G2, yellow), Black Warrior Basin Coal Test (G3, yellow): Mississippi Saline Reservoir Field Test (G4, blue). Yellow Boxes locate potential coupled-use sites found during the regional survey.

Figure 2-1. Location of Coupled-Use Formations in SECARB Region.

Further from the western margin of the Appalachian Mountains, a thick apron of younger sediments (principally shales, sandstones, and the occasional limestone of Cretaceous through Tertiary age) skirt the mountains from the Atlantic Coastal Plain of Virginia and the Carolinas through eastern Georgia and then extend westward into southern Alabama. Although widely distributed along the Atlantic seaboard, these sediments do not approach the thicknesses needed for our applications until one gets to southeastern Georgia. Westward (into Mississippi and eastern Louisiana), this apron merges into a massively thick wedge of sediments deposited in the Mississippi River embayment and along the margins of the Gulf Coast (Figure 2-2).

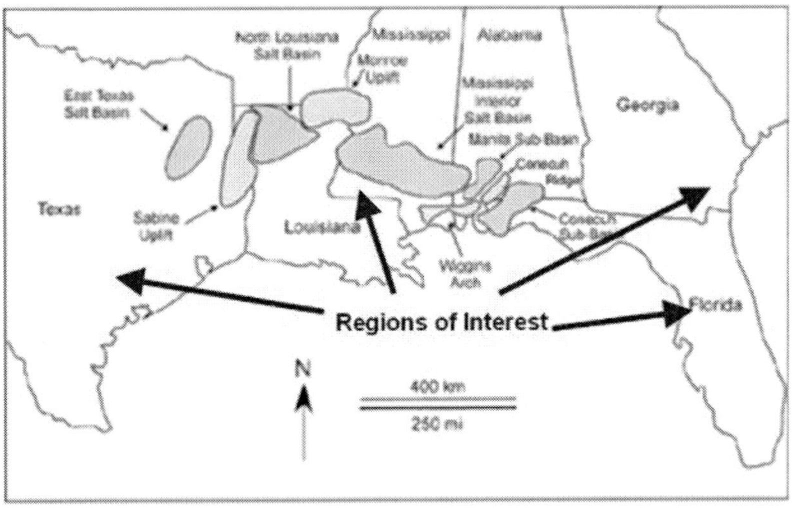

Adapted from Mancini and Goddard 2007.

Figure 2-2. Distribution of Interior Salt Basins in Southeastern United States.

In general, these sediments dip gently southward, and the individual units also usually thicken southward. Because of this southward dip, older rocks (of middle to late Cretaceous age) are exposed at the surface fifty or a hundred miles inland, but are deeply buried by Tertiary (and Recent) sediments along the current Gulf coastline. When the burial depth is great enough to be favorable, the shales (siltstones, mudstones, etc.) can act as migration barriers, while the more porous sandstone units become attractive reservoirs for CO_2 disposal. South of the sediment apron, in Florida, sediments washing down from the Appalachians and interior continent had less influence, and the sedimentary section contains several thousand feet of Cretaceous limestone, which may also act as sequestration reservoirs if there is an impermeable cap above them.

Particularly relevant for this study are the numerous salt domes found in a pair of broad arcs extending west from the Florida Panhandle to the border between the U.S. and Mexico (Beckman and Williamson, 1990). This reflects that throughout the Mississippi embayment and Texas Gulf Coast, the Tertiary and Cretaceous rocks sit on salt beds. Locally, these formations achieve considerable thickness in a series of "basins" extending northwest from the Florida Panhandle to northeast Texas (Figure 2-2, Mancini and Goddard, 2007). The formation of salt domes localized the occurrence of natural resources (gas, oil, and sulfur), and they provided a source of salt that has increased the salinity levels of shallow (often at depths of less than 1,500 feet) groundwaters over much of the area being surveyed in this analysis.

Because of the high salinity of the water of these geological formations, they may not meet the selection criteria for coupled CO_2 sequestration and water extraction and treatment systems. The central criteria for this coupled use are (1) assessing saline formations at depths greater than about 2,500 feet (800 meters) so as to confine the CO_2 at a density (about 0.6 gram per cubic centimeter) and (2) determining a salinity range of approximately 10,000 to 35,000 parts per million total dissolved salts (ppm – the ratio of the weight of the salts in a brine to the weight of the brine initially) so that the displaced formation waters can be desalinated economically for industrial applications. In fact, the target groundwaters would more correctly be termed "brackish water," rather than actual "brines." A further constraint is that current regulation proposals protect brackish waters with less than 10,000 ppm total dissolved solids (TDS) as potential drinking water sources, which renders formations containing such waters ineligible as sequestration targets.

There has been much careful work by Hovorka (Hovorka 1999, Hovorka et al. 2000) showing that if salinity were not an issue, the CO_2 sequestration resource base in the SECARB region would be substantially larger. Absent, however, are surveys that assess the potential benefits of coupling underground CO_2 sequestration with extraction of the displaced brine for additional uses. Underground CO_2 injection tests are planned for these locations, and thus the modeling activities in this analysis could benefit directly from the tests planned for these sites, or vice versa.

There has been much published information that relates to regional hydrologic and geologic trends that should be applicable to more broadly identifying areas where our coupled-use approach might be applied. These resources may not, however, provide answers regarding site-specific, local hydrologic conditions. There are several explanations for this situation. First, when the levels of TDS exceed several ppm, institutions looking for potable water sources may lose interest and stop drilling. Consequently, water chemistry profiles at the depths considered in this analysis may be difficult to locate. Second, the quality of the different geologic surveys differs from state to state, as does the availably of their information resources. A final point is that the upper depth limit (where salinity is likely to be lowest) is too shallow to benefit from the large amount of information generated during (deeper) petroleum or natural gas exploration activities. Additionally, if other projects do target the shallow depths of interest to this study, it may often be at the margins of a salt dome, which may preclude finding saline formation waters with the TDS range under consideration.

An assessment of this nature is actually a two-step process. The first step involves identifying locations where the local geology and hydrology appear to be permissive for a coupled-use project (e.g., formation depth and water salinity). It may prove beneficial to future work to verify that a potential host formation is overlain by at least one impermeable bed (preferably shale or mudstone, for example, for their ability to act as an effective sealing, caprock formation). Tertiary and Cretaceous stratigraphic units found in the Mississippi embayment illustrate the interlayering of porous groundwater containing units, largely sandstones, and essentially impermeable low-porosity mudstones and shales (SECARB, 2008). Demonstrating confinement in a former natural gas reservoir may be convincing because of the history of trapping natural gas over the years (barring impacts of the wells), but for oil and brine fields, the presence of an upper confining bed needs to be explicitly demonstrated.

The second, and more involved part of the assessment process, involves estimating the capacity of a potential reservoir such that the service life of the coupled-use installation can be established. This requires much site-specific information, which is often challenging to obtain, or the coverage is not sufficient to provide a complete analysis. Thus, the resource size estimates provided are at best a preliminary result, as are the generalized brine chemistries suggested as being appropriate for these sites. Finally, this study does not attempt to model the geochemical impacts of injecting CO_2 into the identified formations, but this would be an appropriate future development to more fully assess the formations' suitability for coupled use.

2.2.1. Assessment of SECARB Field Test Sites

The SECARB regional map (see Figure 2-1) shows four sites being evaluated for their CO_2 sequestration potential. The two yellow marks labeled G2 and G3 are coal bed methane production sites where CO_2 is already being introduced to facilitate methane gas production. The green mark (G1) is at a site where oil was produced. Existing wells are to be used at this site to access lower parts of an oil-containing dome-like structure where brine has already displaced the indigenous oil. The blue mark (G4) is at an area, actually several sites close to each other, where deep, porous sandstone is naturally filled with saline water. It is worth noting that both G1 and G4 sites actually target the same rock unit, the Tuscaloosa Formation, which is a substantial regional water-bearing formation. The key element for any of these sites is their porosity, and the porosity level drives the rate of brine displacement. For the coal bed methane sites (G2 and G3), the mechanism by which coal actually absorbs carbon dioxide (while releasing methane) also greatly drives the rate of potential CO_2 sequestration injection rates. This process displaces no pore fluids; however, a great deal of brine is typically brought to the surface under standard methane production activities. The following discussion summarizes the specific SECARB sites from the perspective of a potential coupled-use application.

2.2.1.1. Cranfield Site (G1)

The Cranfield Site is located in west-central Mississippi on a partly expended oil field where a closed dome-shaped structure in the Tuscaloosa sandstone localized a cap of oil floating on brine. Withdrawal of oil has allowed the brine to move upward in the formation toward the top of the dome. According to NETL (2008b), the field test in the lower Tuscaloosa Formation in the Cranfield unit will be at a depth of 10,300 feet, where the planned injection rates may be between 100,000 and 500,000 tons per year of CO_2. As expected for a site located in an oil field, the NATCARB database provided considerable perspective on local brine chemistry (Table 2-1). The initial assessment indicates that the waters there are too saline to be desalinized economically.

2.2.1.2. Central Appalachian Basin Coal Test (G2)

The Central Appalachian Basin Coal Test project will look to inject 1,000 tons of CO_2 into four coal seams in the Pocahontas Formation and four coal seams in the Lee Formation ranging in depth from 1,600 to 2,200 feet. The project will also include coal bed methane (CBM) recovery, whereas the main objective is to demonstrate the application of unminable coal seams to act as a safe method to store CO_2 (NETL, 2008b). For additional perspective, Nemeth (2005) provides an insightful regional overview.

A compendium provided by the EPA regarding environmental impacts of CBM production in this region is available in EPA (2003a), and a comprehensive discussion of the regional geologic setting and water quality issues can be found in EPA (2003b). Unfortunately, much of the reported water-quality information does not include well depths and offers limited insight in the assessment of water-quality information below the 2,500 foot deep criterion. Milici (2004) provides a good discussion of CBM in the Central Appalachian Basin and some general depth information on coal beds of interest to this analysis (Table 2-2) (the Pocahontas coal seam entries are of particular interest). Note that, as indicated in the SECARB site description (NETL, 2008b), the target formation appears to be present at depths too shallow to meet our minimum 2,500 foot depth criterion.

Table 2-1. TDS Levels for Tuscaloosa Formation Brines Close to the G1 Site

Depth Base, feet	Depth Top, feet	Latitude, decimal degrees	Longitude, decimal degrees	Total Dissolved Solids (TDS), ppm
10286	10273	31.49	-91.45	162599
10257	10249	31.49	-91.45	147035
10705	10623	31.49	-91.03	195558
10705	10703	31.49	-91.03	206650
10258	10230	31.5	-91.49	142553
10320	10312	31.53	-91.19	153939
10362	10330	31.53	-91.18	150995
10319	10284	31.53	-91.17	171666
10366	10330	31.53	-91.16	157025
10364	10294	31.55	-91.17	159725
10552	10550	31.56	-91.01	211066

Data adapted from the NATCARB (2008).

Table 2-2. Methane in Coal

Coal Bed	County	Depth (feet)	CH_4 (SCF/ton)
Jawbone	Dickenson	678	278.3
Jawbone	Dickenson	680	278.3
Jawbone	Dickenson	431	281.5
Jawbone	Dickenson	431	156.8
Pocahontas	Buchanan	1,430	435.1
Pocahontas	Buchanan	1,518	463.9
Pocahontas	Buchanan	2,143	339.1
Pocahontas	Buchanan	1,737	348.7
Pocahontas	Buchanan	1,845	351.9

Depth of Methane Containing Coals in Virginia and West Virginia.
Source: DOE, 2003.

This location is the one of the four initially listed in the SECARB region with the least information on local groundwater chemistry. However, the NATCARB database offers well location information relating to the Pocahontas formation in West Virginia (with two entries for Kentucky). Figure 2-3 represents water quality in the Pocahontas Formation over much of southwestern West Virginia, while Figure 2-4 abstracts the data for just the counties specifically selected for Phase II SECARB study sites. Two conclusions can be reached from this evaluation: (1) from a *data analysis perspective,* based on the West Virginia data, the water quality (at relevant depths) in these formations will likely be too saline for the purposes of the coupled-use treatment and (2) the *modeling approach* in this analysis will not be able to be effectively applied because it was unable to quantify the *size of the resource*.

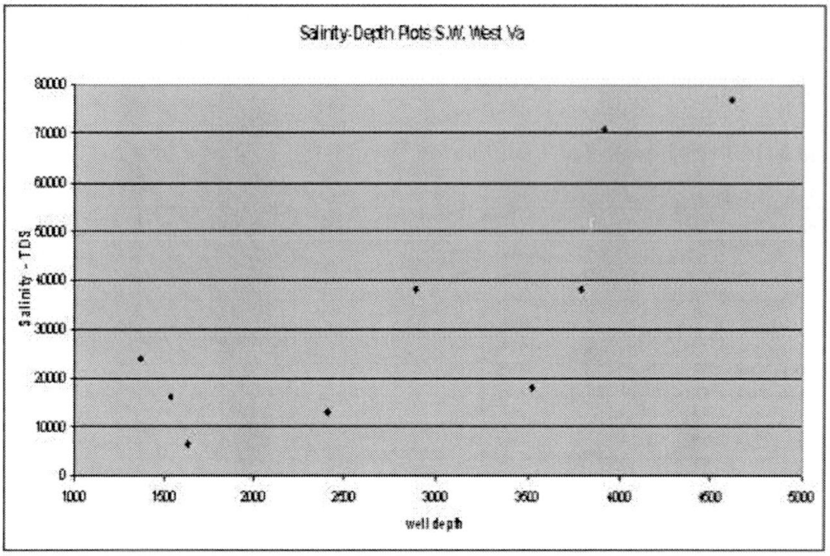

Note: Data from the NATCARB (2008) database, which had no well listings in the designated parts of either Kentucky or Virginia, well depth in feet.

Figure 2-3. Water Chemistry in Wells in SECARB G2 Area of West Virginia.

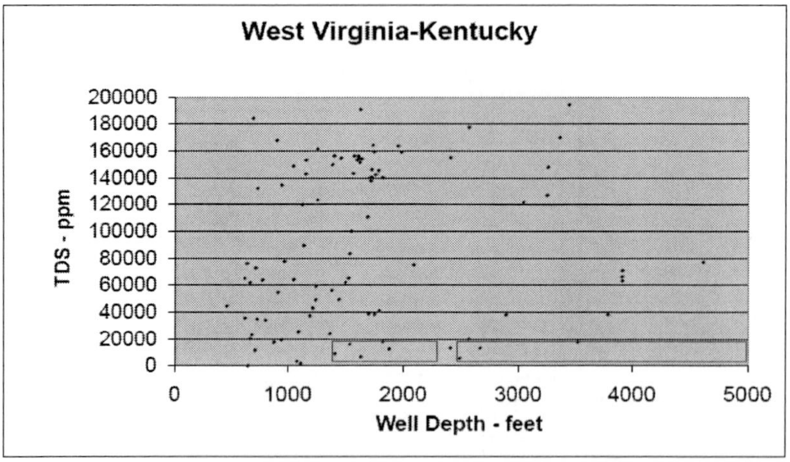

Note: There were only two wells in eastern Kentucky; data from the NATCARB (2008) database. Red box outlines conditions defined in our search for coupled-use sites; green box shows CO_2 injection parameters for coal bed methane sites as presented in NETL, 2008b.

Figure 2-4. Broad Survey of Well Water Chemistry from Southwestern West Virginia and Eastern Kentucky.

2.2.1.3. Black Warrior Basin Coal Test (G3)

The core objectives of the Black Warrior test are to assess CBM reservoirs for their ability to sequester CO_2, while at the same time looking to determine the injectivity rates possible within this region. The test will take place in three coal seams with an injection of 1,000 tons of CO_2 at a depth of approximately 1,500 to 2,500 feet below the surface (NETL, 2008b).

For this analysis, the Black Warrior Basin is a much more attractive candidate than the G2 site because of the substantial database already developed, which includes local groundwater chemistry and the CO_2 sequestration potential of the basin. Additionally, it is also a much larger feature than the G2 site (formally, the basin extends westward from central Alabama into eastern Mississippi). Murrie et al. (1976) provides a good regional overview of the regional geology, and Ortiz et al. (1993), Pashin et al. (1991), Pashin and Hinkle (1997), Pashin and Payton (2005), and EPA (2004) also provide additional context to the regional issues. The basin has a complex structure, and not all the coal seams are at sufficient depth to make them attractive CO_2 sequestration targets. However, there are many coal seams, and it is often reported that good quality water may be encountered while producing methane (EPA, 2004).

There are regional differences in water quality in the basin that can be explained by the fact that in the southwest the coal lies beneath a cover of Cretaceous sands and shales, while in the northeast, the aquifer is "unconfined" (e.g., communicates readily with the surface). In the unconfined portion of the formation, fresh surface waters discharge into the aquifer where the up-turned edges of the coal beds reach the surface. This hydrologic head, in turn, forces relatively low TDS ($NaHCO_3$-rich) waters to considerable depths (Pashin, 2007; EPA, 2004). In this setting, CO_2 sequestration might be somewhat problematic as the system would have to rely on the chemical affinity between coal and carbon dioxide to trap injected gas because there is no overlying impermeable barrier to confine the CO_2.

An additional complication is that the overall TDS level may actually be so low, even below 2,500 feet, that groundwater could be considered as a potential drinking water source (less than 10,000 mg/kg TDS), and by the constraints on this analysis proposed by the EPA (2008), is potentially unavailable for CO_2 storage. In the southwestern part of the Black Warrior Basin, however, the older coal beds (Upper Pottsville Formation of Pennsylvanian age at around 300 million years old) are covered by a cap of younger Cretaceous sands and clays. Predictably, the salinity of the groundwaters rises to levels apparently in excess of 30,000 ppm in places, and NaCl becomes the dominant solute (Table 2-3) (Pashin, 2007; O'Neil et al., 1992).

To summarize, considering both depth and TDS constraints leads to the conclusion that the Black Warrior Basin would be a viable site to explore coupled-use models, but only in the southwest (and deepest) part of the basin. This area is only a small fraction of the whole basin, roughly an area 25 miles long by 10 miles wide, and it may also have the highest-salinity waters. It is also noteworthy that this part of the basin has seen little development activity, which may either mean that it is a resource waiting to be exploited or that access would be more difficult or expensive than for the shallower potential CBM locations in other parts of the basin. Finally, as with the G2 site, there is a clear need to clarify some of the criteria for site selection when CBM properties are the target, since much useful ground might otherwise be excluded based on an inappropriate depth criterion (compare red and green boxes in Figure 2-4).

Table 2-3. Comparison of Water Quality in the Two Parts of Black Warrior Basin

Parameter	Units	Produced Water-Southwest	Produced Water- Northeast
Specific Conductance	µS/cm	27,600–33,800	3,000–6,900
Turbidity	JTU	2–10	0–20
Dissolved Oxygen	ppm	4.2–9.8	1.4–8.2
Hardness	ppm	1,700–2,700	35–130
Nitrate as N	ppm	0–8.6	0–1.2
Ammonia as N	ppm	4.4–7.4	0–0.61
Phosphate as PO_4	ppm	0–4.6	–
TDS	ppm	16,700–21,200	1,900–4,300
BOD	ppm	<1.0–61	–
pH	s.u.	7.7–8.0	7.3–8.5
Bicarbonate	ppm	170–380	670–810
Carbonate	ppm	0	0–74
Calcium	ppm	460–740	7.8–41
Magnesium,	ppm	130–200	3.4–15
Sodium	ppm	5,900–7,200	540–1,800
Potassium	ppm	6.3–20	1.9–3.3
Sulfate	ppm	ND	0–110
Chloride	ppm	9,900–13,000	480–2,000
Silica	ppm	1.9–7.2	–
Iron	µg/L	0–30	10–410
Manganese	µg/L	160–440	0–130

Adapted from: O'Neil et al., 1992.

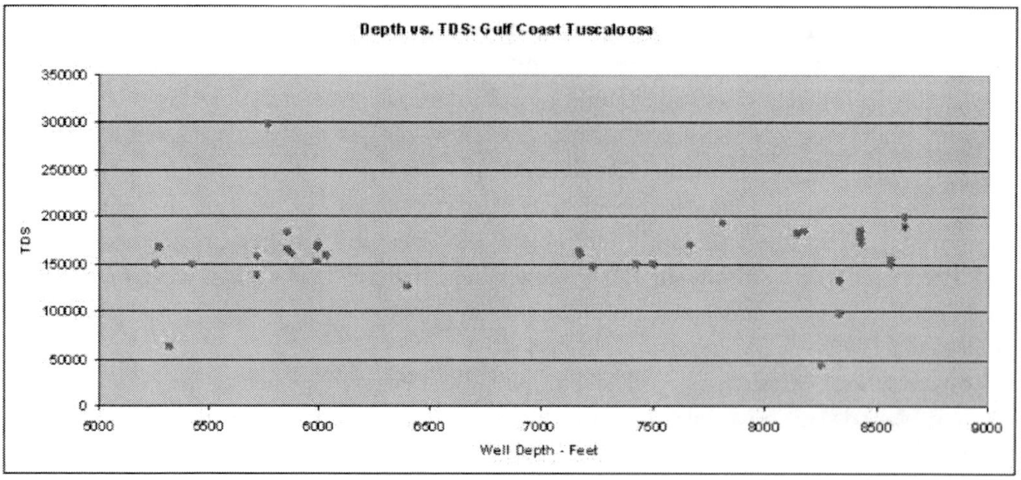

Data Source: NATCARB (2008) brine chemistry database.

Figure 2-5. Distribution of Well Water Salinity (ppm TDS) from Tuscaloosa Formation along the Gulf Coast from Florida to Mississippi.

2.2.1.4. Saline Reservoir Field Test: Mississippi Test Site (G4)

The Mississippi test site will involve injection of 3,000 tons of CO_2 at a depth of approximately 8,600 feet into the Lower Tuscaloosa Massive Sand Unit (NETL, 2008b). This test site differs from the others in that three potential locations are being considered, spanning a length of the Gulf Coast from the Florida Panhandle west to south-central Mississippi. To cover this region, a NATCARB search (Figure 2-5) was made for wells into the Tuscaloosa Formation inside a region extending from 30.7 to 31.2 E. latitude and -87.0 to -89.0 longitude. In this region, well depth may apparently have little systematic impact on salinity, and typical brines have around 150,000 ppm TDS. This is much greater than an initial 20,000 ppm TDS upper cutoff limit that is appropriate for our coupled-use modeling approach.

2.3. Regional Assessment of SECARB Brackish Water Resources

The previous section evaluated the four SECARB pilot test sites for coupled-use applications because of their potential for establishing informative interfaces with other CO_2 sequestration programs. The initial evaluation only targeted four relatively small plots in an area that comprises only a portion of the entire U.S. These considerations suggested that a reasonable approach would be to broaden our search objectives. Although much of the SECARBdesignated area was subsequently ruled out based on regional geologic conditions, two potential coupled-use sites were identified with relative confidence, and a third possibility was noted. The following discussion focuses on characterizing regions with positive attributes, rather than describing in detail why various areas were not chosen.

The criteria used to select (or discard) various areas were the same as those used in evaluating the four SECARB pilot sites, though a wider variety of information resources had to be accessed. In East Texas, the NatCarb (2008) database and United State Geological Survey's (USGS, 2002) produced-water databases provided the basis site evaluation. Figure 2-1 indicates that there are other potentially interesting areas for which the NATCARB database contains no data. A second source of regional groundwater information was the USGS water resources atlas, from which many of the figures in the following discussion were taken (USGS, 1999). However, the intent of the USGS effort was to define potential drinking water resources, so the water quality cut-off limit was often just a few parts per million TDS. For this study, it was assumed that if a water source of a few ppm TDS was documented at depths near the 2,500 foot depth, then a coupled-use resource would likely be found somewhat deeper at the same location. Finally, because of the broad geographic sweep of this search, it was also occasionally necessary to fill in information gaps using site-specific public records, such as local environmental surveys for permitting waste-water injection wells and information from state geologic surveys (where available). With sparse supporting documentation in many areas, the task of assessing the size of a potential coupled-use resource and then assigning a representative water quality to the site may return highly speculative results.

Ultimately, it was possible to identify two areas (south-central Texas and inland central Florida) where a good case could be made for the presence of coupled-use formations. A third area (the South Georgia Basin) also looks promising from a depth standpoint; but no wells penetrate the relevant saline formations, so salinity remains largely uncharacterized. Searches over large areas along the Gulf Coast (west of the Florida Panhandle) and up into the

Mississippi embayment proved unproductive for reasons already outlined in relation to the G1 and G4 SECARB pilot study sites. The next sections describe these three potential sites in more detail.

2.3.1. South Georgia Basin

The geology along the Carolina coast is typical of much of the Eastern seaboard within the confines of SECARB regional designation. Throughout the region, a basement of crystalline rocks (related to the formation of the Appalachian Mountains) is covered with a blanket of sediments that typically thicken eastward out under the Atlantic Ocean. Unfortunately, along most of the coast, the porous sandstones apparently do not reach a sufficient depth until far off the coast. Hence, the issue of groundwater chemistry was moot and not researched further (Smyth et al., 2007[3]).

The one exception to this generality is a feature known as the South Georgia Basin, which is actually nothing more than a down-warped part of the general thick accumulation of Tertiary-Cretaceous sediments (mixed sandstone, shale, and mudstone) that covers the Atlantic seaboard, sweeps around the southern toe of the Appalachians, continues up into the Mississippi embayment, and finally trends southwest along the Gulf Coast to beyond the Mexican border. What makes this collection of sediments more attractive than those on the Gulf Coast is that east of Alabama, the underlying salt basins are absent. Hence, no salt domes penetrate the younger strata, and the chances are greater of finding formations at depths that contain only modestly saline waters. On the coast in east Georgia, the 10,000 ppm TDS boundary appears to be around depths as great as 4,000 feet, but westward at the Alabama/Georgia border, the salinity boundary lies at a depth of just 2,500 feet and is even shallower further west (USGS, 1999). However, the initial research was unable to locate sufficient data indicating how far south of the 10,000 ppm TDS limit the 20,000–35,000 ppm TDS boundary might be or any record of water chemistry from well depths in the 2,400–4,000 foot deep range. Additional insights from the USGS (1999) indicate that in the central Florida section, the mapped information also seems to confirm that sediment depths and water salinities could be conducive to coupled-use applications.

2.3.2. Central Florida (Lower Cedar Keys/Lawson Formations)

The second potential coupled-use site identified is situated in central Florida north of Lake Okeechobee. In much of central Florida, the bedrock geology determines the hydrogeology and is comprised of layers of sedimentary rocks, which tilt gently down to the south. The abundance of limestone in this stratigraphic section reflects that during Cretaceous and Tertiary times, this region was typically a shallow off-shore marine bank that largely escaped the influence of (sand and clay-rich) sediments washing off the North American continent. Thus, the dominant building block for the sediments was calcium carbonate (calcite and aragonite) shells and skeletons that subsequently turned into limestone. Limestone permeability is inherently low, but calcite readily dissolved to the point where hairline fractures widened to open channels, and very high storage capacities may be seen. At about 2,000 feet over the middle of the peninsula, at the base of the confining unit that separates the upper and lower Florida Aquifer system, there is a seawater/freshwater interface. This is defined near the boundary between Cretaceous and overlying Tertiary (Florida Aquifer) sediments (Meyer, 1989).

The main (and uppermost) regional water-bearing rock unit is the "Florida Aquifer." However, in spite of the generality of the nomenclature, the term actually refers to just the thick sequence of near-surface Tertiary rocks and not the whole section of water-containing sedimentary rocks underlying Florida. The boundary between the Tertiary Florida Aquifer and the water-bearing (Cretaceous) units beneath (e.g., the Cedar Keys Formation and Lawson Formation) is located in the middle of the Cedar Keys Formation. Thus, the Cedar Keys is confusingly referred to as being of both Tertiary and Cretaceous age. However, it is particularly important that the middle part of the Cedar Keys is relatively impermeable and likely to confine CO_2 injected into underlying (Cretaceous) formations. Hovorka et al. (2000) reports:

> Central Florida is the focus of this regional assessment because the lower Cedar Keys and Lawson Dolomites appear to contain a laterally extensive porous zone that is overlain by an anhydrite-dolomite sequence in the middle Cedar Keys Formation that is about 700 ft thick, forming an effective top seal (J. Haberfeld, Florida Department of Environmental Protection, personal communication, 2000).
>
> – Hovorka et al., 2000: p.1.

Additionally, they report that *laterally extensive porous zones* being referenced are locally called *boulder zones* and are unit cavernous intervals leached out by circulating groundwaters and locally can have very high storage capacities (Hovorka et al., 2000, p. 1). These features exist beneath the Florida Aquifer in the Lower Cretaceous Pine Key and Lawson Formations and in "Tertiary Cedar Keys and Avon Park Formations" (e.g., in the lower aquifer subdivision of the Florida system) (Hovorka et al., 2000, p. 1).

A note of caution, however, can be found in the USGS groundwater atlas (USGS, 1999: Figure 45), which indicates that boulder zones are found only south of a line trending northeast from Fort Meyers to Cape Canaveral. It also appears that the boulder zones in the (Tertiary) Florida Aquifer system are not well enough confined to be serviceable for CO_2 sequestration:

> The Boulder Zones of the lower Floridan Aquifer are not presented in the GIS because they are generally too shallow, and the confining horizon within the middle and upper Oldsmar and lower Avon Park Formations generally has too high a leakage rate to be expected to retain gases (J. Haberfeld, Florida Department of Environmental Protection, personal communication, 2000), particularly along fracture zones...."
>
> – Hovorka et al., 2000: p.1.

Less is known of conditions in the aquifer immediately below the Florida Aquifer (e.g., separated from the Florida Aquifer by the middle Cedar Keys Formation), but the prognosis is guardedly optimistic as Hovorka et al. (page 1) states, "In general, however, there is currently very little information available. Nonetheless, Florida is a rapidly developing area, and the Lawson and Cedar Keys Dolomites are potential reservoirs for CO_2 sequestration."

Generally, all the water-bearing formations along the Florida coast suffer from salt-water intrusion to some degree depending on local groundwater pumping practices. This raises the TDS even at shallow depths to levels that may not be easy to desalinate. However, for some parts of inland central Florida, the situation may be more promising, but less so toward the southern end of the peninsula. Along a line from Naples to Fort Lauderdale, the salinity even in the (Tertiary) Florida Aquifer is too high to be of use at the 2,500 foot depth limit. Hovorka et al. (2000) supports this view and indicates (Figure 2-6, Table 2-4, well locations unknown) that below 1,700 feet, the water becomes more saline, which will increase the cost for treating extracted water for coupled-use applications and that the waters are typically of the sodium-chloride-sulfate type (Table 2-4).

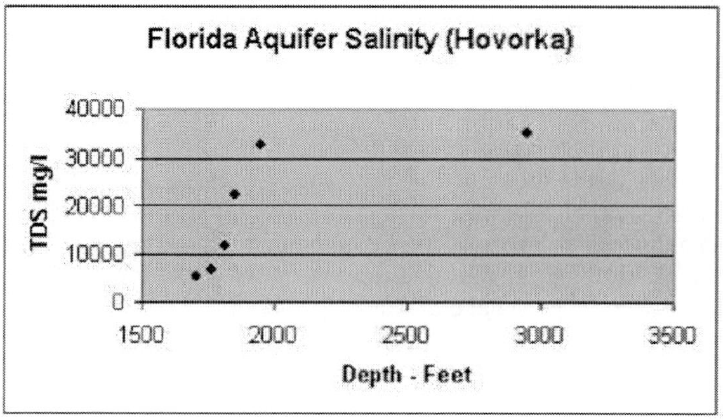

From Hovorka et al., 2000.

Figure 2-6. Salinity Values for Cedar Keys and Lawson Formations, South Florida.

Table 2-4. Water Chemistry for Brines Recovered from the Cedar Keys and Lawson Formations, South Florida

Temperature (degrees C)	Depth (meters)	TDS (ppm)	CA (ppm)	MG (ppm)	NA (ppm)	K (ppm)	HCO3 (ppm)	SO4 (ppm)	CL (ppm)	pH	F (ppm)	(SIO$_2$ (ppm)
22	1705	5690	171	218	1600	59	172	568	2950	8	1.1	11
22.2	1765	6830	203	262	1940	710	168	592	3650	7.9	1.1	10
22.2	1810	11700	276	415	3490	127	170	800	6500	7.9	1.2	10
21.7	1850	22500	405	755	6880	252	162	1440	12600	7.8	1.3	8.5
21	1950	32500	500	1200	9860	380	164	2280	18100	7.8	1.4	6.1
16	2947	35000	428	1300	10800	415	142	2660	19300	7.6	1.4	4.8

Data Compiled From Hovorka et al., 2000; reference cited Vernon, R. O., 1970.

This general picture of the depth-salinity relationships in south Florida is further confirmed by Meyer et al. (1989) in an area extending south from the town of Stuart to the tip of the Florida peninsula and inland as far as Lake Okeechobee (as it relates to the Florida Aquifer system) (see Figure 2-6). Aside from the similarity between the two datasets, what is particularly interesting about the data reported by Meyer et al. (1989) is the sharp break in the salinity/depth plot again around the boundary between the upper and lower parts of the Florida Aquifer, suggesting that the two parts may be isolated from one another. This suggests that, at least locally, there may be a "cap" in the Florida Aquifer, which is not consistent with the view advanced by Hovorka et al. (2000) (see earlier quote) that the Florida Aquifer could not effectively contain CO_2. However, this is a region where the salinity in the lower aquifer is too high for a coupled-use application. As a consequence, this observation would only be useful in situations where water was withdrawn from a different (shallower) formation than that receiving any CO_2.

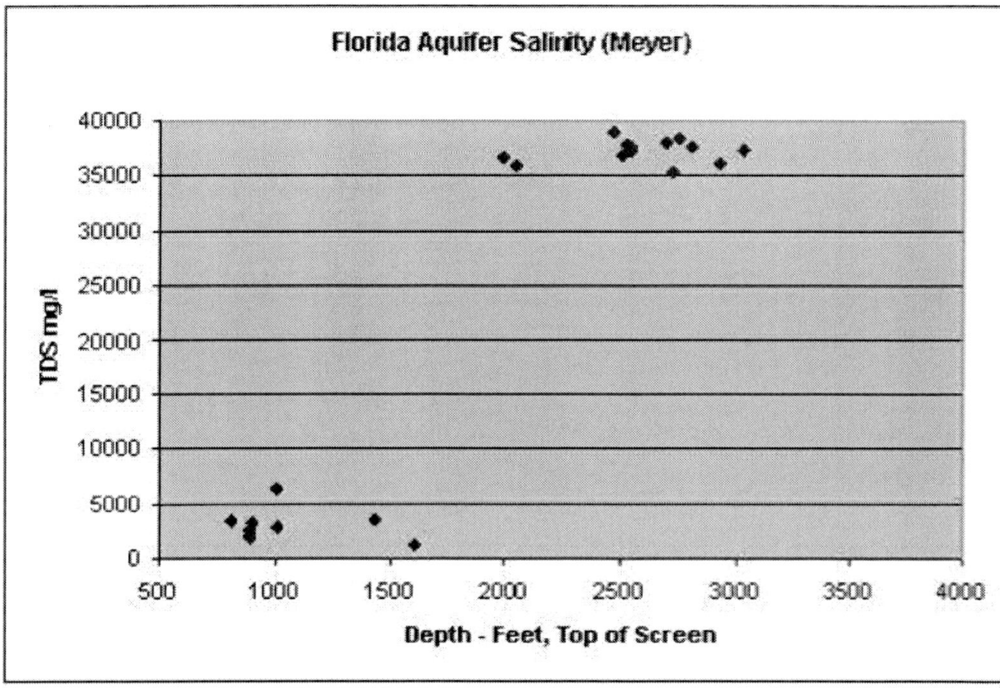

Data Source: Meyer, F.W., 1989: Table 4.

Figure 2-7. Water Chemistry in Florida Aquifer Southeast of Lake Okeechobee.

Even higher salinities are reported for still deeper formation waters in south Florida. Meyer (1989) summarizes data on oil field brines produced from the Lower Cretaceous Sunniland Limestone (beneath the Cedar Key and Lawson dolomite) that indicate salinities in the 200,000 ppm TDS range. An additional regional analysis is provided by Hovorka et al. (2000), indicating TDS values within the 130,000 – 150,000 ppm range. The NATCARB database indicates that wells in south-central Florida are related to pre-Cretaceous strata and also indicated TDS levels far in excess of the 35,000 ppm limit.

Maps from the USGS groundwater atlas (USGS, 1999) indicate that the base of the Florida Aquifer system remains below the 2,500 foot depth criterion until north of Polk County. Additionally, the salinity in the base of the Florida Aquifer system decreases northward to values compatible with a potential coupled-use application. Confinement issues in the Florida Aquifer still remain, but the low salinities at the bottom of the Tertiary aquifer might also indicate a similar situation beneath the confining unit in the middle of the Cedar Keys Formation.

Not much is known about the detailed hydrogeochemistry of deep water-bearing formations north of Lake Okeechobee, but a few isolated waste injection wells have been drilled in Polk County, Florida (Spechler and Kroening, 2006; O'Reilly et al., 2002; Navoy, 1986). These reports confirm that down to at least 2,300 feet, the water is still only moderately saline (Spechler and Kroening, 2006). This suggests that a possible coupled-use brine reservoir may underlie a large area of central Florida. However, issues such as the presence of an adequate upper confining bed and possible updip migration of a buoyant CO_2 plume still need to be evaluated, and more information needs to be gathered and verified about regional water quality.

2.3.3. South-Central Texas Resource

Hovorka et al., (2000) and the USGS (1999) documented the hydrologic resources in Texas to the point that it becomes apparent that the salinity of the Claiborne-Upper Wilcox Groups became an area of interest quickly. Combining the information from these sources, a potential coupled-use site may be found inland roughly 100 miles from the south Texas Coast.

Further discussion of the regional hydrology, and confirmation of a mildly brackish brine reservoir at sufficient depth, was found in documents prepared for the Texas Water Development Board (TWDB, 2003; Dutton et al., 2003; Kelley et al., 2004). The cross sections presented in a manual of brackish groundwater resources prepared for the state (TWDB, 2003) were particularly helpful in defining the extent of the brackish water resource in the Claiborne-Upper Wilcox Formations. Well data and geologic information were compiled to produce four geologic cross sections that show brackish water would be encountered at 2,500 feet. However, these sources also indicated that high-salinity water could also be encountered at a depth of just 1,000 feet. Consequently, not far southwest within this sediment package, the coupled-use resource may no longer exist. Furthermore, the presence of salt domes to the north yields shallow groundwaters below 2,500 feet that are too saline to meet coupled-use criteria.

Additionally, the USGS Produced Water Database (2002) contains many chemical analyses on waters obtained from the Wilcox and (overlying) Claiborne Groups. These data were first sorted by TDS levels. Approximately 21% of the wells in the database for this Texas formation qualified by having TDS levels between 10,000 and 35,000 ppm, as well as having depths greater than or equal to 2,500 feet. Wells that were too shallow, or for which no depth was listed, were also eliminated, leaving the tabulation of water chemistry that appears in Appendix 1: Table A1.

The data analysis showed that Wilcox-Claiborne Formation water chemistries were highly variable. This is particularly true for the proportions of chloride and bicarbonate and the presence (or absence) of sulfate. This reflects both the absence of inputs from concentrated NaCl brine sources (imposed by using an initial upper 20,000 – 35,000 ppm TDS cutoff limit) and the fact that sulfate may be consumed by bacterial action (in conjunction with organic matter) to produce bicarbonate (and sulfide, which typically is removed by forming the mineral iron pyrite, FeS_2). Hence, high-carbonate waters are also typically depleted in sulfate. This rule, however, may not be absolute, since formation waters may initially have had variable sulfate levels (reflecting, among other things, differing access to interbedded layers of gypsum in the sedimentary rocks), and encountered different amounts of organic matter relative to the amounts of sulfate initially present in the water.

2.4. Resource Size Evaluation

Three areas were identified with potential for coupled-use applications: the Carrizo-WilcoxClaiborne brackish water-bearing formations in Texas, beneath the middle Cedar Keys Formation in central Florida, and the Black Warrior Basin. The Texas site has sufficient stratigraphic and hydrological control to allow for an initial volume estimate. In contrast, only a small amount of hydrologic and geochemical information could be found on the Florida site, so any volume estimate may be only a rough, initial estimate that would require further verification. The Black Warrior Basin is unique in that an exhaustive assessment of storage capacity has already been made, which needed only to be screened for our program-specific applications.

The Texas site (the Claiborne-Upper Wilcox hydrologic system, also referred to as the CarrizoWilcox system) lies in an elongated band parallel to the Gulf Coast but inland about 100 miles. The work by the USGS (1999) indicated that the length along a northeast-southwest trend for the deep, low salinity part of the saline formation is roughly 250 miles. Their work also suggested that the width of the band lying between 3000 and 10,000 ppm TDS is on the order of 15 miles. A 15 by 250 mile area equates to 3,750 square miles, and a vertical thickness for the saline formation would be at least 1,000 feet (TWDB, 2003). The porosity of the Carrizo sand is 30– 40% (Pearson and White, 1967). The volume in a box 250 miles long, 15 miles wide, and 1,000 feet thick with porosity of 35% is, in total, 1.04×10^{12} cubic meters of pore space.

Somewhat better map coverage is available for the middle part of the area of interest (Dutton et al., 2003; Kelley et al., 2004), though again one has to work with contours that do not actually define the size of a region lying between 10,000 and 20,000 – 35,000 ppm TDS. In this case, the area over a linear distance of 220 miles between the 10,000 and 50,000 ppm TDS contours is approximately 3,700 square miles. Operationally, this is virtually the same area as was defined above, but it is likely that only about half would be underlain by brackish waters with less than 20,000 – 35,000 ppm TDS. Hence, there is actually a difference of a factor of roughly two in the two estimates. Part of the discrepancy arises because, with improved maps, it became evident that the width of the band of moderately brackish water swells and shrinks by a considerable factor along the trend. Thus, the 15 mile width estimate used previously may be a little too wide.

However, it is also evident that the area evaluated in the Dutton et al. (2003) and Kelley et al. (2004) studies encompasses only a part of the resource. The brackish water manual prepared for the TWDB (2003) suggests a southwest continuation into the part of Texas where the sequestration resource was originally identified based on the USGS groundwater atlas (USGS, 1999). Additional conformation of this southwest extension comes from evaluating the geographic distribution of wells from the USGS Produced Water Database (2002) (e.g., wells meeting both the depth and TDS criteria for the Wilcox Formation). Most of these wells are located along a trend leading northeast from Lavaca County toward Houston. However, there is also a cluster of qualifying wells to the southwest (e.g., in an area located inland from Corpus Christi, roughly half way to Laredo). This confirms that the size of the resource exceeds what would be inferred from the map coverage provided by Dutton et al. (2003) and Kelley et al. (2004).

In summary, the initial 15 mile band may have been somewhat wide, which was determined based on better maps of the middle part of the area that was evaluated. However, it is also evident that sequestration opportunities are available southwest of the areas considered by Dutton et al. (2003) and Kelley et al. (2004). Hence, it is likely that the original estimate of 1.04×10^{12} cubic meters is a reasonable initial measure of the size of the Carrizo-Wilcox (and Claiborne) sequestration resource. However, it is noteworthy that when CO2 is injected into porous sandstone, not all of the brine can be displaced. When calculating the capacity, it is presumed (for consistency with other estimates) that 100% displacement could be achieved. In reality, a number much lower than this (e.g., perhaps closer to 10%) affecting the efficiency of the brine displacement process. The TWDB (2003) summarizes the finding well in the following paragraphs.

> The Carrizo-Wilcox aquifer consists of hydrologically connected, Tertiary age, interbedded sands, clays, silts, and some discontinuous lignite beds of the Wilcox Group and the overlying massive sands of the Carrizo Sand. Overlying the aquifer are the shales and clays of the Reklaw Formation in East and Central Texas, and the sands and clays of the Bigford Formation in South Texas, which serve as the confining units over the Carrizo-Wilcox. The thickness of the Carrizo-Wilcox varies widely across the state. In South Texas, saturated thickness in the outcrop is generally less than 100 feet, and in down-dip sections reaches only a maximum of 500 feet. In South Central Texas, the aquifer thickens significantly, with as much as 700 feet of saturated thickness in the outcrop areas and up to 2,000 feet in down-dip portions of the aquifer. In Central Texas, the aquifer thins somewhat, with outcrop saturated thickness of less than 500 feet, and thickness down-dip of 1,000 to 1,500 feet. In East Texas, saturated thickness in the outcrop is generally less than 500 feet and down-dip thickness is a maximum of 1,000 feet. Four general cross sections for the Carrizo-Wilcox are shown....
>
> Aquifer properties of the Carrizo-Wilcox aquifer vary across the state. In South Texas, transmissivities are relatively low, ranging from 5,000 to 10,000 gpd/ft. In South Central and Central Texas, transmissivities are much higher, and can range from 50,000 to 150,000 gpd/ft. In East Texas transmissivities are high, ranging from 30,000 to 50,000 gpd/ft. In general, the interbedded sands and clays of the Wilcox are less permeable than those of the Carrizo aquifer. Coefficients of storage for the Carrizo-Wilcox aquifer range from about 0.0001 to 0.001 in the artesian areas, and unconfined specific yields range from 0.05 to 0.30. Well yields from the more transmissive sections of the Carrizo-Wilcox can be thousands of gallons per minute. These characteristics are from the fresh sections of the aquifer, and the aquifer is generally expected to be slightly less prolific in the down-dip areas where slightly- to moderately-saline water is found.
>
> – TWDB, 2003, pp. 50-51.

At the Florida site (Figure 2-8), far less information is available on which to base a CO_2 sequestration capacity estimate. In fact, very little is known of the geochemistry of the lower Cedar Keys and Lawson Formation waters, but parameters relevant to estimating reservoir size have been considered by Hovorka et al. (2000). Porosities in the range of 24–28% are suggested, as is a formation thickness of 1,260 feet. However, the capacity calculations in this analysis use only a third of this thickness since, with increasing depth, it is likely that higher salinity levels will be encountered.

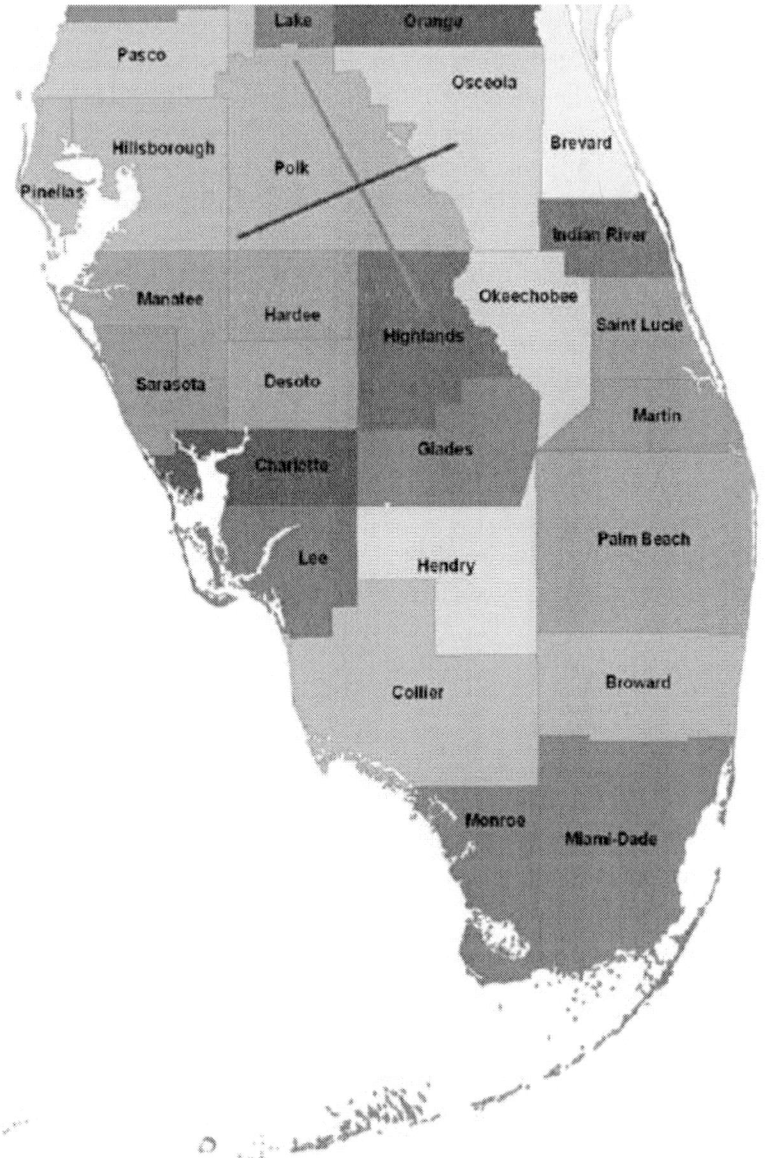

Figure 2-8. East-West and North-South Extent of Probable South-Florida CO_2 Sequestration Resource Area.

To obtain an estimate of the "north-south" extent of the area, the analysis focuses on the northern border of Polk County. The depth to the base of the Florida Aquifer system (and hence the top of the confining unit for the underlying lower Cedar keys–Lawson Aquifer) is about the minimum depth acceptable for CO_2 injection, 2,500 feet (see Figure 2-8). Further south, the formation is even deeper, meeting the depth criteria. At the relevant depths from Lake Okeechobee south (but in the Florida Aquifer in this case), the salinity may begin to reach levels considered too high for coupled-use applications, roughly 35,000 ppm TDS (Meyer, 1989: Table 4). Sprinkle (1989), however, suggests that the salinity (again in the lower Florida Aquifer) generally decreases northward to around 1,000 ppm TDS in northern

Polk County, and the report presented by Spechler and Kroening (2006) seems to confirm low salinity at depth, at least in Polk County and in the lower Florida Aquifer. Hence, the initial assumption was made that the salinity in the upper part of the underlying Cedar Keys–Lawson Aquifer might also be below approximately 20,000 ppm TDS for locations north of the middle of Highlands County, but further verification would be necessary to validate this assumption. In the "east-west" direction, the width of the box (assuming a similar correspondence between the Florida and underlying aquifer systems) would be a line extending from the eastern border to Polk County to the middle of Osceola County, again based on a reading of the salinity contours illustrated in Sprinkle (1989).

This defines a rectangular area that is roughly 60 by 80 miles. Using porosities and thicknesses discussed previously, this leads to a very preliminary reservoir pore volume calculation of 4.3×10^{11} cubic meters. As previously mentioned, the degree to which the brine can be displaced by CO_2 is subject to debate. In this case, however, the displacement may be larger than in the Texas example, since the solution channels formed in carbonate aquifers are typically both large and interconnected (e.g., such formations are very "transmissive").

Finally, the Black Warrior Basin illustrates a third assessment for an initial storage capacity estimate. Our earlier survey suggested that using depth criteria designed for porous sandstones resulted in only a relatively small area (a 10 by 25 mile square) being identified as suitable. However, the volume calculation methodology for CBM sites is different from that for sandstone, and fortunately, the basin as a whole has already been assessed in detail.

With this estimate in hand, the question is whether water-quality factors would substantially diminish the size of the apparent resource. In fact, water quality is relatively good throughout much of the basin, and the main concern might actually be that some fraction of the deep water might still be protected as a potential drinking water source (e.g., less than 10,000 ppm TDS currently, but 3,000 ppm TDS in the published study). For current purposes, it is likely sufficient to use the Pashin et al. (2003) report as a baseline; converting five trillion cubic feet of CO_2 (see below) translates into about 2.6×10^{11} kg of carbon dioxide (TCF is trillion cubic feet of gas at 60 °F and one atmosphere pressure).

> Applying the geologic screening model indicates that significant parts of the coalbed methane fairway are not accessible because of fault zones, coal mines, coal reserves, and formation water with TDS content less than 3,000 ppm. Excluding these areas leaves a sequestration potential of 60 MMst (4.7 Tcf), which is equivalent to 1.9 years of emissions. Therefore, if about 10 percent of the flue gas stream from nearby power plants is dedicated to enhanced coalbed methane recovery, a meaningful reduction of CO_2 emissions can be realized for nearly two decades. If the fresh-water restriction were removed for the purposes of CO_2 sequestration, an additional 10 MMst (0.9 Tcf) of CO_2 could feasibly be sequestered. The amount of unswept coalbed methane in the fairway is estimated to be 1.49 Tcf at a pressure of 50psi. Applying the screening model results in an accessible unswept gas resource of 0.44 Tcf. Removal of the fresh-water restriction would elevate this number to 0.57 Tcf. If a recovery factor of 80 percent can be realized, then enhanced recovery activities can result in an 18 percent expansion of coalbed methane reserves in the Black Warrior basin.
>
> – Pashin et al, 2003, p. iv.

Table 2-5 summarizes the capacity estimates described in this section.

Table 2-5. Summary of SECARB Region Coupled-Use Sequestration Capacities

Site	Approximate Storage Capacity
Black Warrior Basin	2.6 x 10^{11} kg of CO_2 (Pashin et al., 2003)
Central Florida	4.3x10^{11} m^3 and 2.6 x 10^{14} kg @ 100% CO_2 Occupancy*
S. Central Texas**	1.04x10^{12} m^3 and 6.2 x 10^{14} kg @ 100% CO_2 Occupancy*

* Modeling needed to assess what fraction of the porosity is actually accessible. **The CO_2 density was assumed to be 0.6 g/cm^3. Although 100% brine displacement is used to compute metrics for the Texas site, a more conservative value for our modeling purposes would be closer to 10% of the capacity in this table.

2.5. Representative Water Chemistries

To complete the coupled-use study, it is necessary not only to assess the size of the sequestration resource, but also to define chemistries for the brine potentially being displaced by the injected gas. Given the challenging processes needed to identify the sites in the first place, it therefore proves difficult to define representative water chemistries. For the central Florida site, the regional literature (particularly O'Reilly et al., 2002, and Navoy, 1986) suggests a sharp increase in salinity near the depth relevant to CO_2 sequestration with chloride concentrations close to 10,000 ppm. Hence, the 1,850 foot depth sample from Hovorka et al. (2000) was selected to represent likely water chemistry from the Florida Aquifer. Table 2-3 contains both shallow- and deep-water chemistries for the Black Warrior Basin, and the range in water chemistries given for the deeper water is likely representative of what would actually be available for coupled-use applications. As noted earlier, the water chemistry in the Wilcox (Carrizo, Claiborne) Group is highly variable. For this reason, the analysis uses the "average chemistry" provided on the first line of data in Table 2-5 initially. Later, if important distinctions become apparent, other selections from the table can be used to address specific issues.

2.6. General Summary

The SECARB area has a complex geologic history and hence diverse hydrology. It follows that locating a suitable coupled-use site will be a challenging task. Only a few areas were determined to be preliminary candidates such that they met both the requisite criteria to hold a porous water-bearing formation while also at sufficient 2,500 foot depth. As summarized in Table 2-6, the Black Warrior Basin CBM site seemed to (marginally) meet the requisite criteria, and a broader regional analysis identified only two (perhaps three) other workable areas. The two regional potential locations with the best documentation are in central Florida and, in particular, the inland site in south-central Texas. Southeastern Georgia (the Georgia Basin) also seems promising, but there are insufficient water-quality data from this region to make a definitive recommendation.

Table 2-6. Representative Formation Water Chemistry Data

Component	Alabama Black Warrior Basin Low TDS	Alabama Black Warrior Basin High TDS	Florida Cedar Keys/ Lawson	Texas Wilcox/ Claiborne
Sodium *	5,705	7,337	6,880	7308
Magnesium	130	200	775	68
Potassium	6.3	20	252	Not Given
Calcium	460	740	405	475
Bicarbonate	170	390	162	1066
Sulfate	Not Given	Not Given	1440	507
Chloride	9,900	13,000	12600	11310
Silica	1.9	7.2	8.5	Not Given
Iron	0	30	Not Given	Not Given
pH	7.7	8.0	7.8	7.2

*Small adjustments made to literature values for sodium to achieve exact charge balance.

Additionally, porous sandstone and limestone aquifers generally hold more promise than coal bed methane plays, since the available sequestration volumes appear to be much greater.

Beyond site-selection issues, there is also a fair amount of learning to be done regarding sequestration processes in the different rock types. In the limestone present in Florida, the porosity is likely to be in the form of large vugs and cracks, so capillary forces are not likely to impede displacing the brine with CO_2. However, finding the porous zones in limestones at depth can be quite difficult and may require advanced geophysical techniques. Just the opposite is true for porous sandstones found in Texas. The porosity of the rock is likely to be somewhat predictable throughout much of a formation, but capillary forces greatly complicate the process of modeling how much of the brine can be displaced during gas injection. For the true percent of displacement, a test case could be expected to show that the brine displaced by way of CO_2 injection is highly site and condition-specific. The percent of water recovered from the pore volume will greatly affect the engineering, economics, and overall viability of any potential coupled-use site. An additional consideration is whether reservoir sequestration capacity may be improved when water is actively pumped out of the formation. Finally, coal beds (Black Warrior Basin, Alabama) retain CO_2 by multiple mechanisms. In addition to just displacing the brine from existing pores, the coal can incorporate carbon dioxide into the molecular structure of the coal and surrounding strata. Carbon dioxide binds to indigenous methane, which is why CO_2 is already being injected in some regions to facilitate methane production (e.g., without explicitly accounting for any benefit to sequester the CO_2). This process then causes the coal to swell, which affects porosity and the gas gaining access to the bulk of the coal.

A final finding was that there may be many more coupled-use possibilities if the formation receiving the CO_2 does not have to be the same as that supplying the water for desalinization. It has been established for some time (Hovorka et al., 2000, and many others) that the CO_2 sequestration capacity of deep, high TDS, brine-filled formations along the Gulf Coast is immense. Additionally, during the search for coupled-use sites, it was common to find records indicating the presence of relatively shallow (1,000–1,500 feet deep) mildly brackish waters sitting above the deeper high-salinity aquifers. Additional details obtained from the NATCARB database provided this analysis with additional insights into the location of potential brackish water resources in the SECARB region (see Appendix 1 for supplemental details). Hence, it may be relatively straightforward for utilities to find sites where wells of differing depths could meet both the sequestration and (brackish) water supply and CO_2 storage requirements for fossil fuel-fired power plants and other industrial sources of CO_2.

Among all of the sites examined for this geoassessment, the Black Warrior Basin and the Carrizo-Wilcox system met most of the geoassessment criteria. To complete the analysis of potential sites for coupled-used systems, Section 3 examines a preliminary geomodeling of CO_2 injection in the Carrizo-Wilcox system, and Section 4 incorporates the information developed in the initial geoassessment of the four general areas outlined in yellow boxes in Figure 2-1 within the SECARB region and adds the power plants to complete the first-round analysis of locating a candidate site for a more formal geomodeling activity. By choosing this site, the analysis will give a full start-to-finish method that can be followed to determine the geological, spatial, power plant technology, and finally geomodeling components of a full-site characterization and assessment analysis. With this example framework, future sites may be included to determine the high-level information required to formally assess their suitability as a coupled-use site for carbon sequestration and water extraction and treatment.

3. Storage Constraints for CO_2 in Sloping Formations with No Structural Closure: Carrizo-Wilcox Formations of East Texas

Abundant sedimentary wedges exist along the eastern coastal United States, associated with the Triassic-aged opening of the Atlantic Ocean and Gulf of Mexico, which could serve as storage sites for atmospheric CO_2. At relatively large depths, these formations bear saline pore (connate) waters, which could be used in the proposed coupled-use scenarios we propose for coal-fired power plants. As these formations slope gently seaward, they offer no structural "traps" for buoyant CO_2, and thus the question arises as to the fate and transport of CO_2 injected at depth. A central question to be answered is "Can sufficient CO_2 be stored in these sedimentary systems over sufficient time to avoid contact with potable water sources in more shallow aquifers by upwardly mobile carbon dioxide plumes?"

Nicot and colleagues (Nicot, 2008; Nicot et al., 2009) investigated, via single-phase modeling using MODFLOW and CMG-GEM, the influence of injection of CO_2 at industrial-scale rates on updip freshwater supplies in coastal plain aquifers. They showed minimal impact, although localized effects could exist. The limitations of the models permit no inclusion of salinity effects or dissolution of CO_2 into the brine phase, so our study is in part motivated by concerns of saline and aqueous supercritical or gas phase CO_2 intrusion on shallow groundwater aquifers. These authors also assume maximal capillary entry pressure for the overlying seal and, by design, do not allow upward migration through seal bypass systems, which most likely occurs at least to some extent in all gas reservoirs. For example, the upward vertical flow observed in the Central Carrizo-Wilcox calibrated groundwater availability model (Dutton et al., 2003) suggests some amount of upward groundwater flow in our study area. One unknown is how much upward migration of CO_2 as a supercritical phase and/or as a dissolved constituent in brines would be accommodated by this flow system.

To develop a regional scale numerical model of CO_2 injection and storage in Texas Gulf Coast Basin saline formations, we finalized an earth or geologic framework model for the CarrizoWilcox and Queen City Sparta saline formations and began preliminary TOUGH2 simulations of CO_2 injections. The TOUGH2 model is based on a MODFLOW model that forms part of the central Groundwater Availability Model (GAM) prepared for the TWDB by staff from INTERA, Inc., and the Texas Bureau of Economic Geology (Kelley et al., 2004). Figure 3-1 shows the grid of the central model (shown in grey) and a subset, outlined in black, in use for the TOUGH2 simulations. This model runs from the outcrop of the Carrizo-Wilcox (shown as tan strip in the north to southwest portion of the model grid) to the updip limit of the Wilcox growth fault zone (blue line running northeast to southwest), which also marks the transition from hydrostatic to geopressured regions of the Carrizo-Wilcox Group Formations. Choice of model location was also dictated by proximity to coal-fired power plants discussed later in this report. Shown as green lines are contours of salinity (as TDS in ppm) in the deeper portions of the formations. The TOUGH2 simulation scenarios initiate with CO_2 injection within this region and extraction of saline groundwater, tracking both the evolution of CO_2, groundwater salinity and fluid pressure. The light green- yellow dots in Figure 3-1 show locations of groundwater wells in the Carrizo-Wilcox Aquifer system that will limit the extent of updip migration of CO_2 plumes in the simulations (i.e., the simulations are aimed at determining how long CO_2 can remain down-dip of the freshwater zones and if substantial CO_2 migration could take place vertically through the caprock). This assumes that injection would stop once some indication of CO_2 plume migration is detected in the groundwater wells.

The background gray grid is from the Central Carrizo-Wilcox GAM (Dutton et al., 2003); each square denotes one square mile. The blue line denotes the Wilcox Growth Fault boundary and also approximately delineates the western extent of geopressure in the Wilcox Formation. The green contours plot regions of varying salinity in the Carrizo-Wilcox updip of the overpressured region. The purple, red, blue, and yellow dots show most groundwater well locations that produce from the Carrizo-Wilcox (purple indicates wells completed in the Hooper Formation, red indicates Simsboro, blue indicates Calvert Bluff, and yellow, the Carrizo Formation wells), and the brown swath denotes the location of the Wilcox outcrop and recharge area in east Texas. Simsboro (our proposed injection target) and Reklaw (proposed regional caprock) are outlined in blue and burgundy respectively. The purple/black dots show locations of oil and gas exploration and production wells. Total dissolved solids in ppm, if available, are given next to the well locations. The large black dot shows the location of the Fayette Power Project coal-fired power plant, which outputs 11.9 M tonnes of CO_2 per year.

To properly assess the extent to which CO_2 injection influences both pressure, flow regime, and salinity changes in overlying sediments, as well as the potential for CO_2 intrusion, we have initiated both analytical and numerical investigations of CO_2 injection and storage in Carrizo-Wilcox clastic formations. In our numerical models, we use the TOUGH2 family of codes (Pruess et al., 1999; Pruess, 2005), which allow for multiphase CO_2-brine effects but do not permit (in our version) hysteresis in capillary pressure and thus are not equipped to properly address trapping of CO_2 as a residual gas phase. Appendix 2 describes most of the techniques used to model the CO_2 in the Carrizo-Wilcox Formation. The simulations provide guidance for this and future analyses as to the size of the CO_2 plume, size of the formation, and the rate of the plume's migration over time, given a set of geological and geophysical conditions. An analytical model that addresses the updip migration and retention of CO_2 by capillary trapping in sloping formations with no structural closure is given in Appendix 2. Similarly, a TOUGH2 3D numerical model of CO_2 injection and migration coupled with brine migration relevant to the geography and geology shown in Figure 3-1 is presented in Appendix 2.

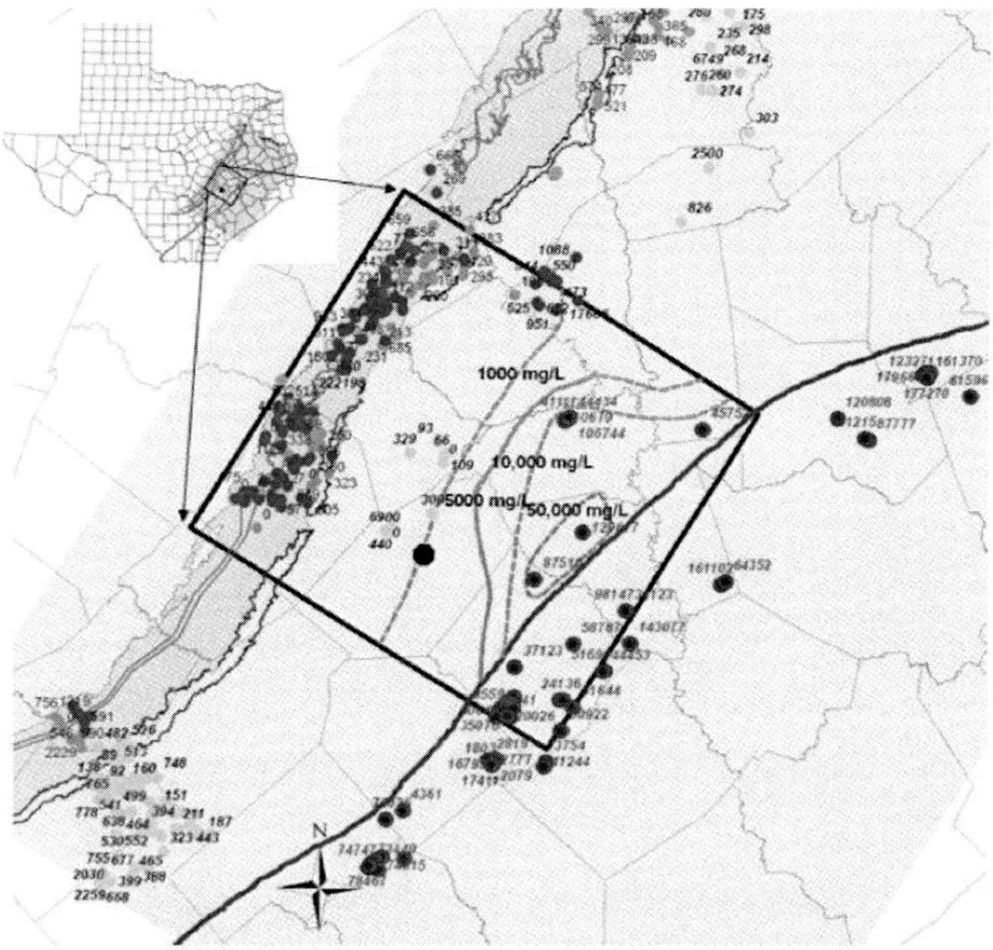

Figure 3-1. Location of TOUGH2 Simulation Domain (Black Rectangle) in Central Gulf Coast Region of East Texas.

4. Regional Analysis: Representative Power Plants and Potential Water Treatment

4.1. Spatial Analysis of Formation and Power Plant Matching

The SECARB region of the U.S. has been evaluated for characteristics promising for coupled-use for simultaneous withdrawal of groundwater for beneficial use coupled with injection replacement with CO_2. Specifically, all minor regions within the major SECARB region have been analyzed and down selected to determine their suitability as a location from a coupled carbon sequestration and saline formation water extraction perspective while accounting for the types and numbers of power plants within those regions.

To address a potential site for coupled use, locations were identified within the SECARB region containing brackish waters that could be extracted for power plant operations while simultaneously providing porous formations where CO_2 might be sequestered long-term underground. The initial assessment outlined the geological formation requirements and then identified representative power plants relatively close to those formations. The specific criteria for identifying potential coupled-use formations are as follows:

- Groundwater must be between 10,000 and 35,000 ppm TDS and
- Formation receiving the CO_2 must be greater than 2,500 feet deep (800 meters) below the surface (Allis et al., 2003, NETL 2008a).

Table 4.1 summarizes the findings regarding the coupled-use of formations for water supply and simultaneous CO_2 storage.

Table 4-1. Suitability of SECARB Pilot Test Sites for Coupled-Use Applications

Site and Type	Depth Criteria Met?	Water Quality Considerations
G1 – Porous sandstone aquifer Cranfield	Yes	Unlikely – Formation waters are very saline
G2 – Coal bed methane Central Appalachian Basin	Yes	Unlikely – Formation waters probably very saline but very little groundwater chemistry is available for this site.
G3 – Coal bed methane Black Warrior Basin	Partly	Meets criteria
G4 – Porous sandstone aquifer, Mississippi Saline Aquifer	Yes	Unlikely – Formation waters very saline

Note: Analysis provided by Jim Krumhansl.

Figure 4-1 (a reproduction of Figure 2-1) shows an expanded view of the SECARB region with the areas highlighted in yellow boxes indicating potential coupled-use locations. It is apparent when viewing the SECARB region that much of the area lacks sufficient well data to conclusively determine coupled-use locales. Based on the available data, the yellow boxed areas are the most promising.

To further refine our study, locations of existing power plants were overlaid with the coupled-use data. Figure 4-2 shows the location of existing power plants within the SECARB region. The southeast part of Texas, the Black Warrior Basin, and a portion of Central Florida are the only locations identified in the initial geoassessment where existing power plants overlap saline formations that may be considered for potential coupled use with CO_2 sequestration. Once again, it should be noted that saline water data in the regions other than Southeast Texas and the Black Warrior Basin are quite limited.

The TDS level in the Black Warrior Basin is in the range of 16,700 to 21,200 ppm, whereas the average TDS level in the southeast Texas region is 15,390 ppm. The remainder of the area is either very saline (>40,000 ppm TDS) or data are very limited and hence no conclusions can be made.

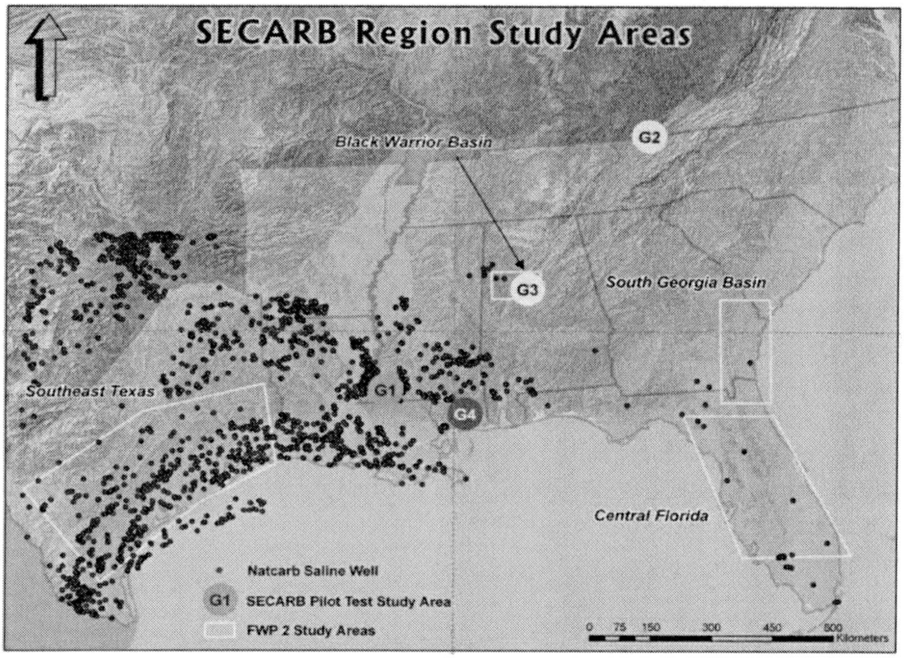

Data Source: NATCARB, 2008; NETL, 2008a, 2008b.

Figure 4-1. Location of Coupled-Use Formations in SECARB Region.

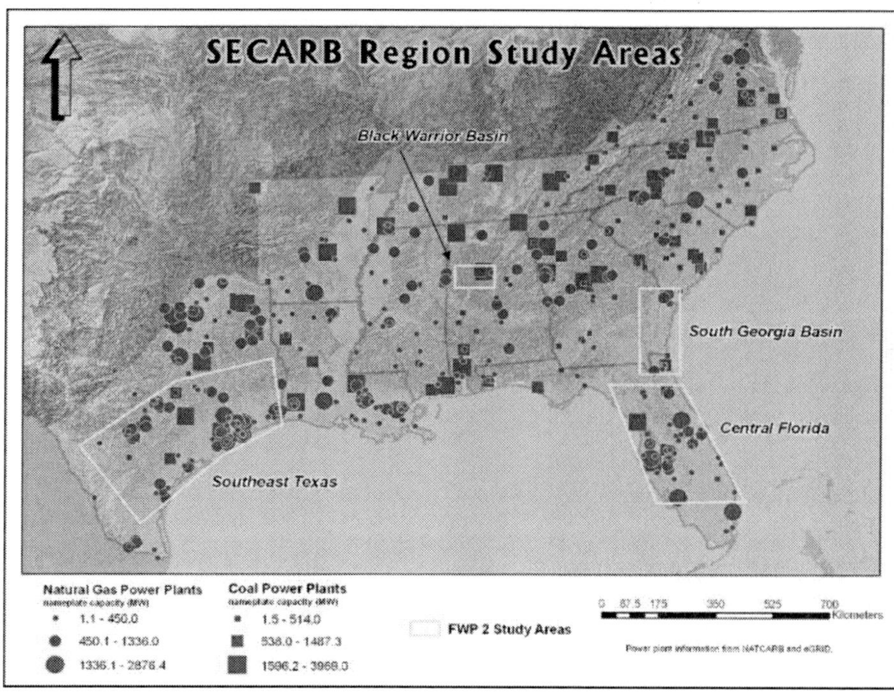

Figure 4-2. SECARB Region – Location of Existing Power Plants.

4.2. Illustrating the Geoassessment

Of the areas identified in Figure 4-2, formation water chemistry in proximity to existing power plants was analyzed in detail. Each figure shows the nameplate capacity of coal and natural gas power plants and nearby well locations with information including well depth and formation water TDS. Some of these potential areas have more wells than others, and in some locations, the wells intersect a formation at a desired depth of 2,500 ft and greater, with a water chemistry of 10,000 to 35,000 ppm TDS. Other areas may also have favorable subsurface geology with the desired depth and chemistry, but there are no wells that can be used to evaluate these formations. Therefore, the analysis is limited to those areas where existing well data were available for screening these four areas.

Within these four areas, the analysis selected the largest power plants with the highest volume of CO_2 emissions within a range of up to 60 miles from what was roughly considered a key location above the formations of interest. It is likely cost prohibitive to consider distances between a potential formation and a power plant that is greater than 60 miles. Table 4.2 shows the power plant information obtained from the Emissions & Generation Resource Integrated Database (eGRID) (EPA, 2009). This information was used to determine the most suitable location for a coupled system with CO_2 injection and brackish water extraction and treatment.

Figure 4-3. Power Plants, Produced Water Wells and CBM Fields – Black Warrior Basin.

4.2.1. Black Warrior Basin

The Black Warrior Basin map shows the outcrop locations of the Pottsville Formation, a potential saline formation for CO_2 sequestration, as well as both nearby coal- and natural-gas–fired power plants. Previous studies identified existing CBM fields with CO_2 capacity at 350 psi (Bcf), and those data are shown in Figure 4-3 with a color chart ranging from green (low capacity) to red (high capacity). Based on the water chemistry from produced water wells that terminate in the Pottsville Formation, a TDS contour estimate of 10,000 ppm is shown on the map. Figure 4-4 shows the CBM capacities with the formation overlay removed for additional clarity.

Narrowing down the nearest power plants to the best formation leaves three power plants just north of existing CBM fields that have been previously analyzed for sequestration by the Alabama Geological Survey. Table 4-2 shows the details of each power plant as well as the distance to the center of the closest CBM field with the largest CO_2 capacity.

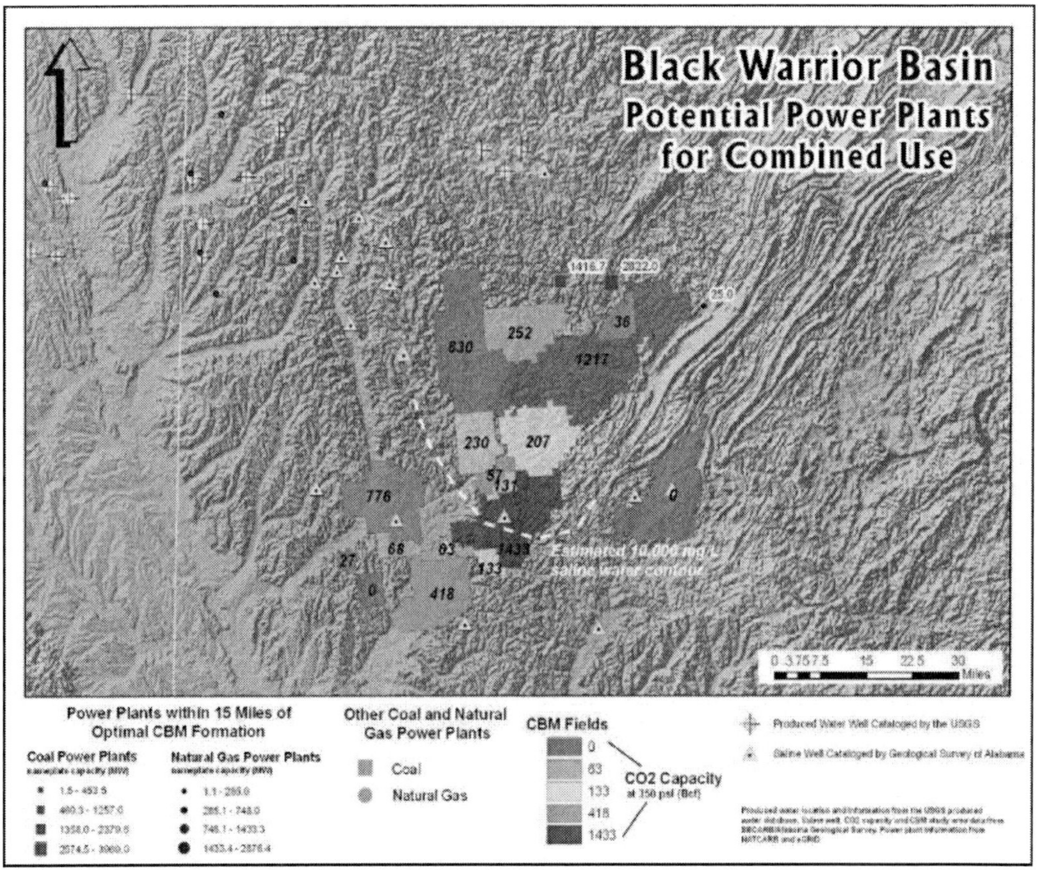

Figure 4-4. Selected Power Plants – Black Warrior Basin.

Table 4-2. Potential Power Plants near Combined-Use Formations – Black Warrior Basin

ORISPL (Plant identifier number)	Power Plant Name	Fuel Type	Capacity Factor	Nameplate Capacity (MW)	Plant Built Date	Distance to CBM field (~miles)	Notes
8	Gorgas	Coal	0.6374	1416.7	See Notes Column	15	5 generators, 1951, 1952, 1956, 1958 and 1972
6002	James H. Miller Jr.	Coal	0.8628	2882.0	See Notes Column	14	4 generators, 1978, 1985, 1989, and 1991
50359	Sloss Industries, Inc.	Coke Oven Gas	0.1863	25.0	1958	22	2 generators

Office of Regulatory Information Systems Plant Location (ORISPL).

4.2.2. Central Florida

Figure 4-5 is a map of Central Florida showing the location of an identified saline formation along with produced water wells, all near existing coal- and natural-gas–fired power plants. The water quality is somewhat saline with all wells above 35,000 ppm.

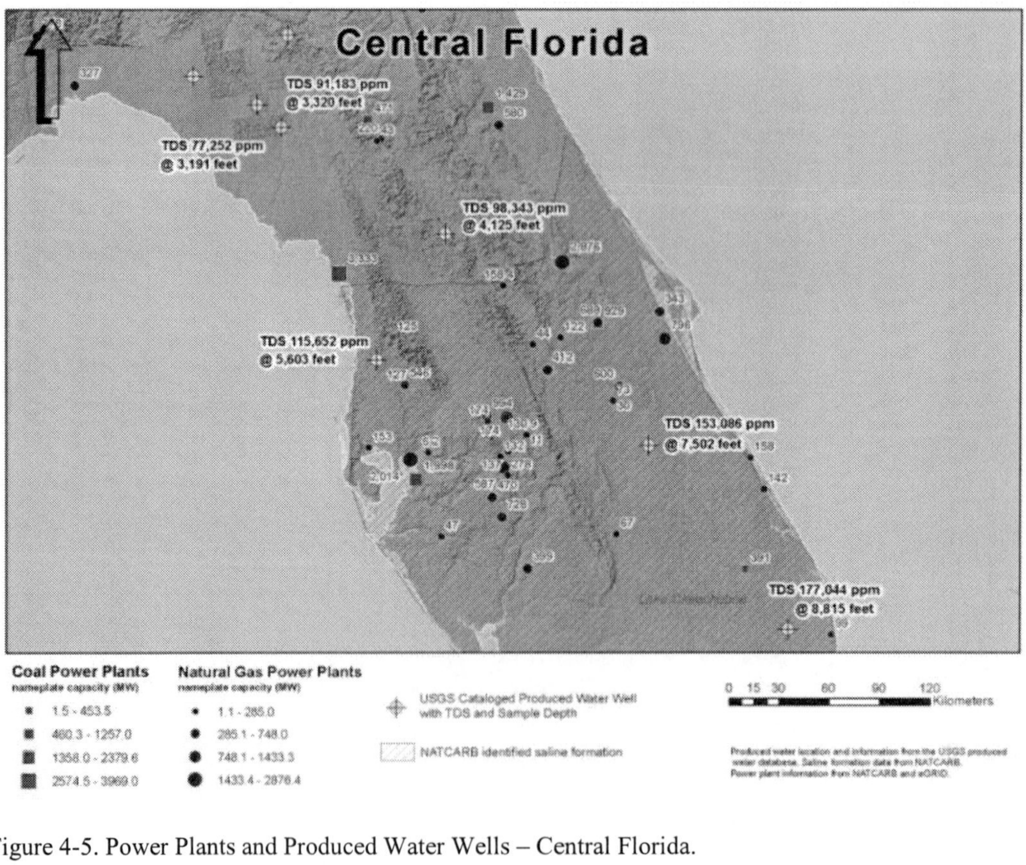

Figure 4-5. Power Plants and Produced Water Wells – Central Florida.

The area in Central Florida with an assumed saline formation that could be utilized is located on the north end of the NATCARB identified saline formation (blue hatched line in Figure 4-6). There is one coal and seven natural gas power plants located in the vicinity. Because there are no well data available in what is considered the best location geographically, a determination on whether this formation meets the coupled-use criteria cannot be made. With regard to available well data, two wells located northwest and southeast and different depths have very high TDS, which is not suitable as determined by the combined-use constraints.

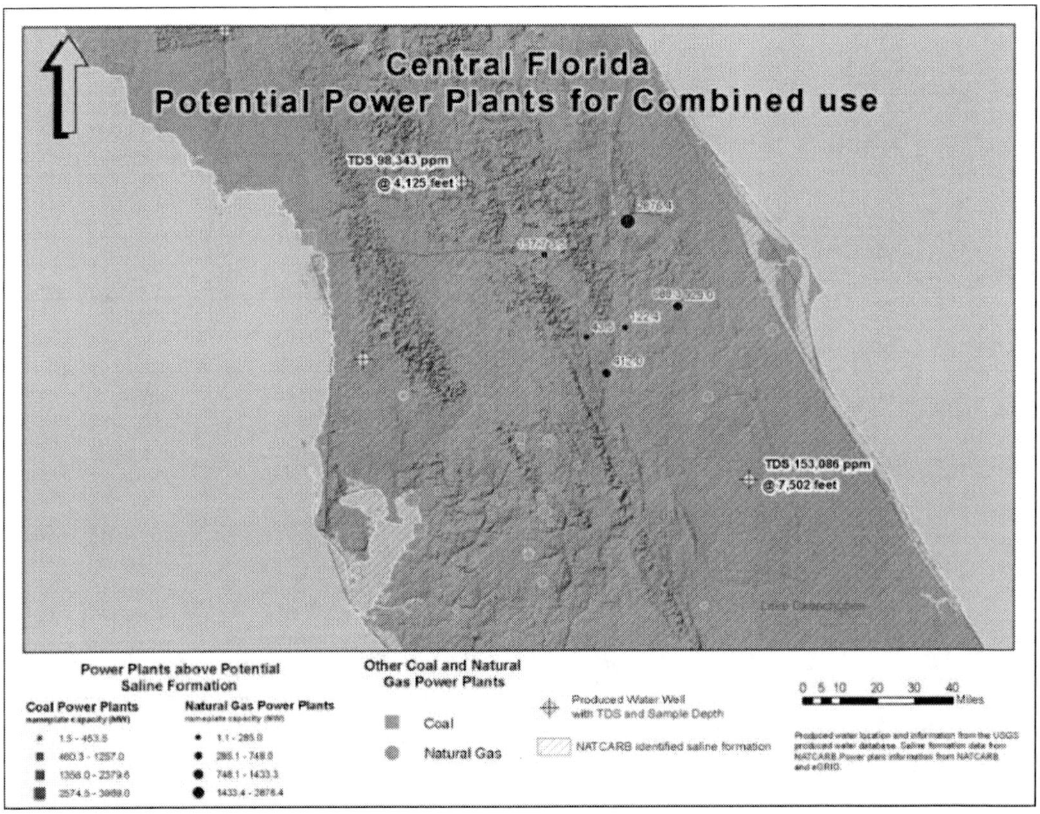

Figure 4-6. Selected Power Plants – Central Florida.

Narrowing down the nearest power plants to the best formation leaves eight power plants located in the northeast of an area identified by NATCARB as having an underlying saline formation. Table 4-3 shows the characteristics of each power plant in this area. Nearby produced water wells to the northwest and southeast have very high TDS values and may not be suitable for combined use. Wells were not identified in the area near the power plants, and therefore, desired water quality at the preferred depth is not available to evaluate.

Table 4-3. Potential Power Plants near Combined-Use Formations – Central Florida

ORISPL	Power Plant Name	Fuel Type	Capacity Factor	Nameplate Capacity (MW)	Plant Built Date	Notes
564	Stanton Energy Center	Coal	0.7684	929.0	See Notes Column	2 generators, 1987 and 1996
54466	Orlando Cogen LP	Gas	0.8042	122.4	1993	
620	Sanford	Gas	0.5314	2876.4	See Notes Column	11 generators, 1959 (oil), 1969, 1974, 2002 (7), and 2003
7238	Cane Island	Gas	0.3906	412.0	See Notes Column	4 generators, 1994, 1995 (2), 2002 (2)
7294	Central Energy Plant	Gas	0.1920	43.5	1989	2 generators, 1987 and 1996
55821	Curtis H Stanton Energy Center	Gas	0.4621	688.3	2003	3 generators
10020	Cutrale Citrus Juices USA	Gas	0.0000	3.5	1987	
54423	Lake Cogen Ltd	Gas	0.3235	157.7	1993	3 generators

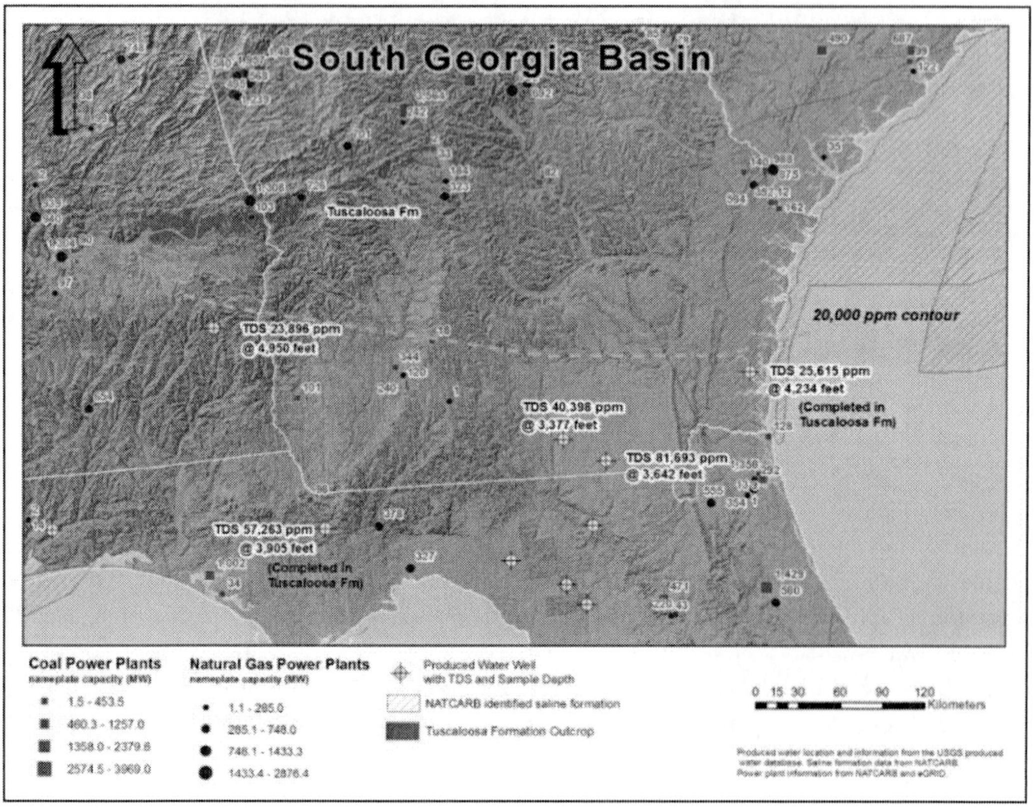

Figure 4-7. Power Plants and Produced Water Wells – South Georgia Basin.

4.2.3. South Georgia Basin

Figure 4-7, the South Georgia Basin map shows produced water wells in the vicinity of existing coal- and natural gas-fired power plants. Water chemistries show a TDS and depth completed in the Tuscaloosa formation near the coast that might fit the criteria for CO_2 sequestration and saline water use. Based on a well located further west in Alabama, there appears to be a rough 20,000 ppm TDS contour that extends west to east in the southern part of Georgia (see Figure 4-7).

The South Georgia Basin has only one known produced water well that intersects a saline formation near the desired depth and TDS. The only known power plants are located south of the well in Florida. The NATCARB database identified a saline formation (blue hatched line) that extends to the Georgia/Florida state line. Six coal power plants and three natural gas power plants were identified with distances between 30 and 60 miles of the produced water well. Power plant details are presented in Table 4-4.

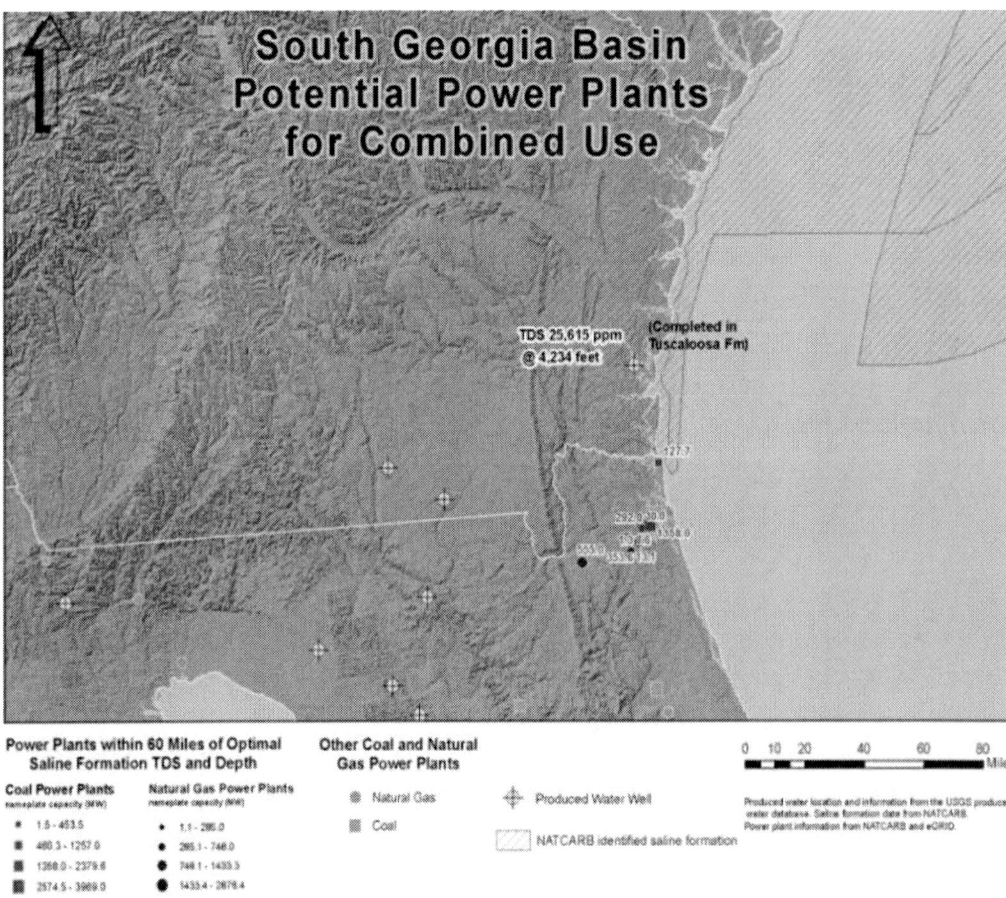

Figure 4-8. Selected Power Plants – South Georgia Basin.

Table 4-4. Potential Power Plants near Combined-Use Formations – South Georgia Basin

ORISPL	Power Plant Name	Fuel Type	Capacity Factor	Nameplate Capacity (MW)	Plant Built Date	Distance to Well in Optimal Formation (miles) approximation	Notes
207	St Johns River Power Park	Coal	0.7311	1358.0	See Notes Column	47	2 generators, 1987 and 1988
10672	Cedar Bay Generating LP	Coal	0.7617	292.0	1984	48	
10202	Jefferson Smurfit Fernandina Beach	Coal	0.5604	127.7	See Notes Column	30	2 generators, 1982 and 1988 (wood waste)
54535	St Vincent's Medical Center	Gas	0.0084	1.3	1991	54	
50803	Stone Container Seminole Mill	Gas	0.3628	30.0	1957	46	
10431	Anheuser Busch Jacksonville	Gas	0.0057	8.6	1987	54	
10008	Baptist Medical Center	Gas	0.4782	13.1	See Notes Column	54	7 generators, 1973 (2), 1972, 1982, 1983, 1986, 1993
7846	Brandy Branch	Gas	0.1645	555.0	See Notes Column	60	4 generators, 2001 (3), 2005
666	J D Kennedy	Gas	0.0187	353.6	See Notes Column	54	4 generators, 1973 (3 oil), 2000

4.2.4. Southeast Texas

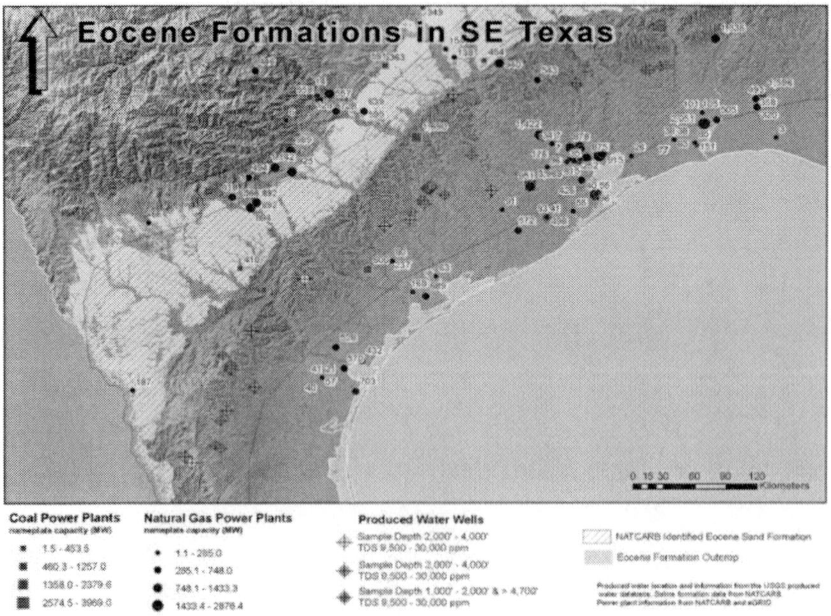

Figure 4-9. Power Plants and Produced Water Wells – Southeast Texas.

For a more detailed assessment of power plants to produced water wells, Figure 4-10 shows the saline formations (Eocene Sands) that the NATCARB database identified in a blue hatched pattern. All the produced water wells that intercept that formation at the desired TDS and depth create a linear feature that trends from southwest to northeast. A buffer of that linear feature was created to capture the existing coal and natural gas plants within 30 miles of the buffer center. Other power plants not captured by the buffer are grayed out. The power plants of interest for this study area include two coal-fired and 13 natural gas plants. Table 4-5 shows the identifying information for each power plant.

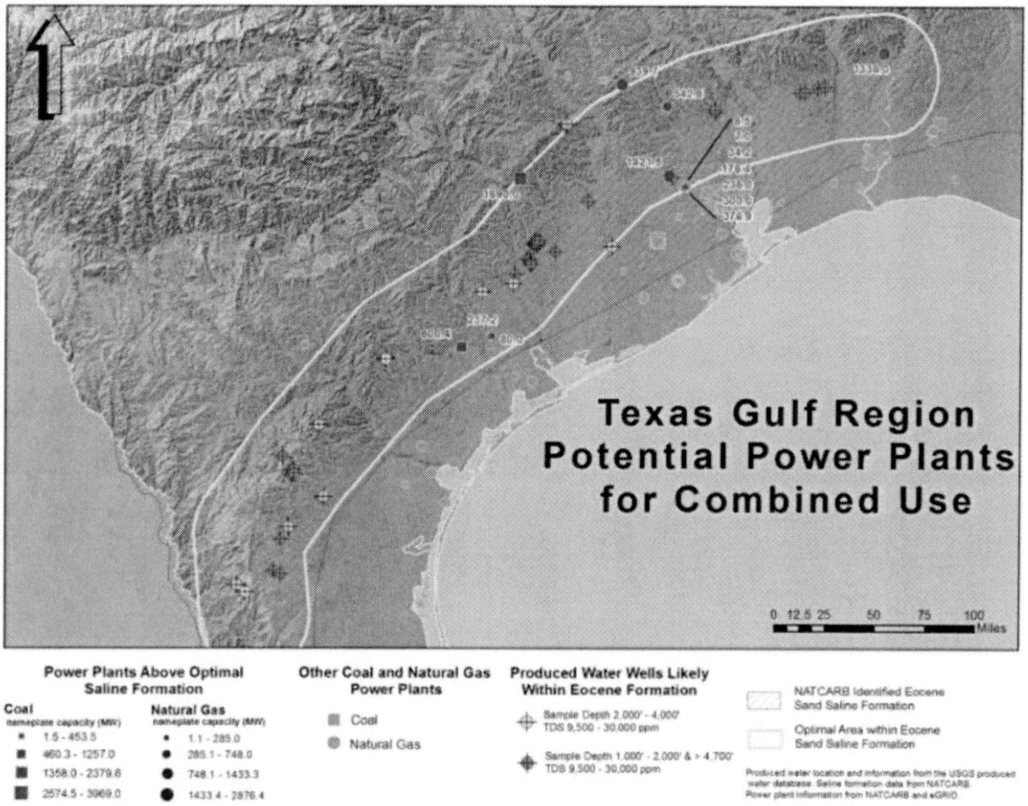

Figure 4-10. Power Plants and Produced Water Wells – Southeast Texas.

4.2.5. Selection of Study Area for Coupled Use

Southeast Texas was chosen for study of the economic feasibility of coupled use because of the high concentration of power plants, available saline formations meeting the TDS, and depth criteria. For the water treatment analysis, two coal and two natural gas power plants were chosen because of their proximity to the saline formation. These power plants are highlighted in Table 4-5. Figure 4-11 shows the relative location of these four power plants to nearby produced water wells. Both shallow and deep waters were assessed to evaluate the cost trade-offs of pumping lower TDS waters from greater depth vs. higher TDS waters from shallower depths. Table 4-6 shows the well depths, TDS values, and the distances from the selected power plants. It is assumed that any water treatment installation would be located with a power plant.

Table 4-5. Potential Power Plants near Coupled-Use Formations – Southeast Texas

ORISPL	Power Plant Name	Fuel Type	Capacity Factor	Nameplate Capacity (MW)	Plant Built Date	Notes
6178	Coleto Creek	Coal	0.9703	600.4	1980	
6179	Fayette Power Project	Coal	0.7470	1690.0	See Notes Column	3 generators, 1979, 1980, and 1988.
3469	T H Wharton	Gas	0.1309	1421.5	See Notes Column	18 generators with span of 1960 to 1975 for on-line date. Most in mid 1970s.
55062	Tenaska Frontier Generation Station	Gas	0.5609	939.7	2000	4 generators, all 2000
52012	Valero Refining Texas Houston	Gas	0.5026	34.2	1990	2 generators, both 1990
10790	Victoria Texas Plant	Gas	0.5749	80.0	1987	
10298	Bayou Cogen Plant	Gas	0.6977	300.0	See Notes Column	4 generators. 2, 1984 and 2, 1985
55358	Cottonwood Energy Project	Gas	0.2250	1336.0	2003	8 generators, all 2003
55365	Exelon LaPorte Generating Station	Gas	0.0397	236.0	2001	4 generators, all 2001
10692	ExxonMobil Baytown Turbine	Gas	0.7621	376.9	See Notes Column	3 generators in 1989, 1 in 1997 and 1 in 2004 (5 total)
3457	Lewis Creek	Gas	0.4039	542.8	See Notes Column	2 generators, 1 in 1970 and 1 in 1971
10638	Pasadena	Gas	0.1276	6.5	1985	
50054	Rice University	Gas	0.5017	7.0	See Notes Column	2 generators, 1 in 1986 and 1 in 1989
3631	Sam Rayburn	Gas	0.2405	237.2	See Notes Column	Mix of natural gas and diesel (9). 1964 (2), 1965 (1), 2003 (4), 1991 (2 diesel).
7325	San Jacinto Steam Electric Station	Gas	0.7959	176.4	1995	2 generators

Table 4-6. Selected Power Plant Water Treatment Characteristics

Power Plant Name	County	Fuel Type	Well Depth (feet)	TDS Level (ppm)	Wellhead Distance to Power Plant (miles)
Coleto Creek	Goliad	Coal	3,631	13,900	34
Coleto Creek	Goliad	Coal	2,700	27,562	30
Fayette Power Project	Fayette	Coal	3,500	11,660	32
Fayette Power Project	Fayette	Coal	2,800	22,848	34
T. H. Wharton	Harris	Gas	2,800	19,568	1.8
Sam Rayburn	Victoria	Gas	2,700	27,562	23.2
Sam Rayburn	Victoria	Gas	3,631	13,900	55

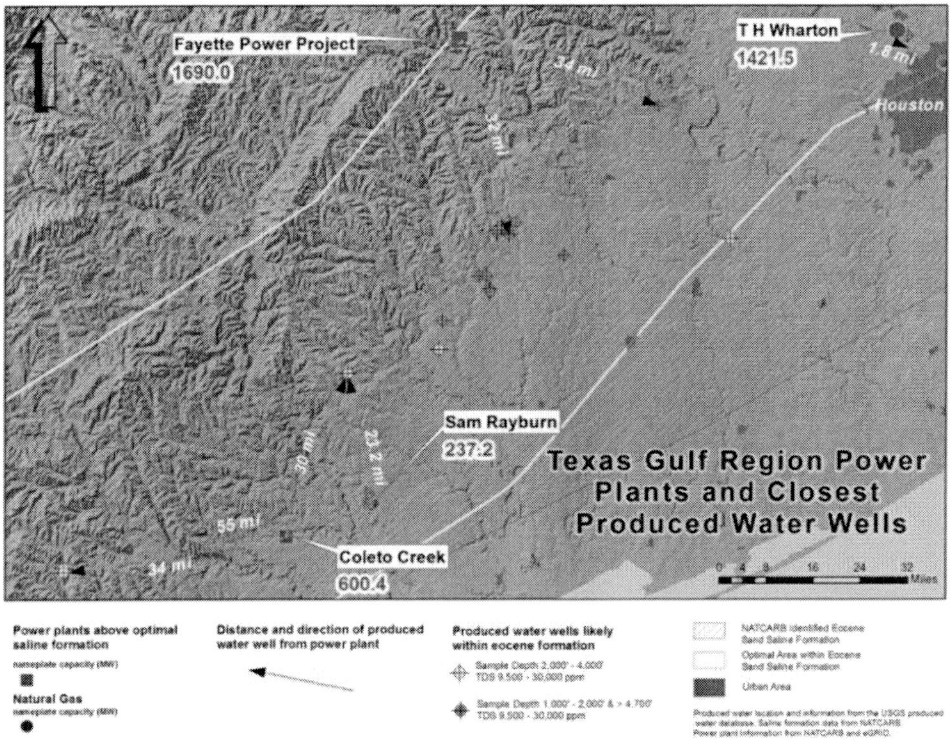

Figure 4-11. Geographic Distribution of Four Power Plants Highlighted in Table 4-5.

4.3. Evaluation of Other Water Demands

To fully appreciate the potential value of the saline water displaced by the sequestered carbon, one must consider it in the context of growing water needs of the region. To assist with this analysis, we make use of the 2007 State Water Plan for Texas and related resources (TWDB 2002, 2007). The state water plan is based on a bottom-up consensus-driven approach to water planning that involved 16 regional water planning groups (Figure 4-12). According to Texas Water Development Board guidelines, each regional planning group reviewed water use projections and water availability estimates in dry or drought-of-record conditions to develop a water budget. Where shortfalls in the budget were identified, the planning groups recommended water management strategies to meet the need.

The Carrizo-Wilcox Aquifer represents a viable water resource in four of the state's 16 water planning regions; regions h, k, l, and n. Projected water demands from 2010 to 2060 for each of the regions are given in Table 4-7. Total water use is projected to increase from 4.2 to 6.3 million acre-feet per year, a 50% increase. The largest increase occurs in the municipal sector where water use grows from 1.5 to 3 million acre-feet per year, a 200% increase. Other large water users include irrigation (1.5 million acre-feet in 2000) and manufacturing (0.8 million acre-feet in 2000). The fourth largest user of water across these planning regions is the thermoelectric sector at 0.2 million acre-feet per year in 2000. However, the thermoelectric sector is also the fastest growing, projected to experience a 249% increase by 2060 (0.57 million acre-feet in 2060). Ultimately, these projections suggest that the thermoelectric sector is in fierce competition for "new" water supplies from the municipal and manufacturing sectors (note that irrigation demand declines by about 19% between 2010 and 2060).

As part of the water planning process, each planning region also assessed available water supplies under drought conditions. These supplies were compared to the demands of Table 4-7 in efforts to identify potential shortfalls. Table 4-8 shows the projected shortfalls under drought for each of the four regions in 2010 and 2060. The projected shortfalls are on the order of 25 to 33% of the total water demand. In other words, water use would need to be curtailed by a quarter to a third in the event of a strong drought. Disruption of water supply is an issue of particular concern to the electric power industry.

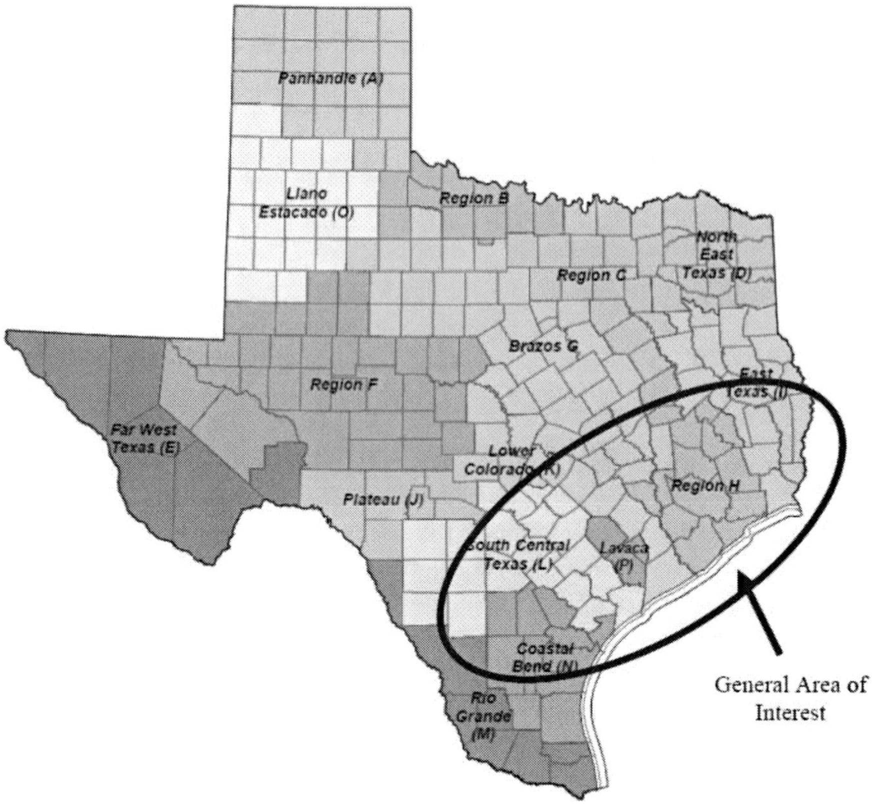

Adapted from TWDB, 2007.

Figure 4-12. Location of the 16 Regional Water Planning Areas in Texas.

In response to the projected shortfalls, each region proposed various water development projects to mitigate the effects of drought. Projects include such actions as interbasin transfers, expanding current well fields, water reuse, conservation, and new reservoir construction. Table 4-9 provides the total projected improvement in water supply and accompanying cost for all projects aggregated at the planning region level. Additionally, the cost per acre-foot of water yielded is also provided. Values range from a low of $415/acre-foot to a high of $7,219/acrefoot. These projected costs provide a rough measure of the value of water in the planning region.

Another interesting point is that a number of the proposed water projects involve utilization of desalinated seawater and brackish aquifer water. In total, 137,000 acre-feet of such projects are proposed.

While the focus has largely been on saline formations in this analysis, it is also important to recognize that the use of alternative water supplies for use at power plants has been examined in the larger research community. Examples include evaluating mine pool water from underground coal mines (Veil et al., 2003; Veil and Puder, 2006); the feasibility of using coal bed methane (CBM) produced water and finally reclaimed water.

Treated municipal and industrial waste water, often referred to as reclaimed water, may be an attractive alternative (Veil, 2007). The report by Veil (2007) states that there are approximately 57 power plants within 16 states currently using reclaimed water for power plant cooling, and some are also using it for air pollution control equipment. Additionally, the report states that nearly all regulatory requirements related to the use of reclaimed water are found at the state level.

As stated clearly by Veil:

> Reclaimed water must meet at least secondary treatment standards. Often the state agency that authorizes reclaimed water use for cooling will require tertiary treatment or some form of filtration and disinfection. In addition to treating the reclaimed water for public health and environmental reasons, operators often treat for process reasons. Reclaimed water typically contains chemical parameters that may create problems of scaling, corrosion, biofouling, or stress cracking of some metal surfaces.
>
> – Veil, 2007, p. 28.

Thus, potentially utilizing so called reclaimed water for alternative uses (such as cooling water in power plants) presents a different set of treatment challenges to those of extracting deep saline formation waters, but it is a viable combination given the appropriate conditions.

Table 4-7. Projected Water Demands 2020–2060 for the Four Planning Regions That Utilize the Carrizo-Wilcox Aquifer as a Water Source

Projected Water Demands (Acre-feet)								
Region	Category	2000	2010	2020	2030	2040	2050	2060
H	Irrigation	464,330	450,175	438,257	433,686	430,930	430,930	430,930
	Livestock	12,228	12,228	12,228	12,228	12,228	12,228	12,228
	Manufacturing	628,025	722,873	783,835	836,597	886,668	927,860	950,102
	Mining	49,473	57,043	60,782	63,053	65,285	67,501	69,457
	Municipal	850,091	980,544	1,116,660	1,253,607	1,391,710	1,552,375	1,732,608
	Steam Electric	83,262	91,231	112,334	131,332	154,491	182,720	217,132
	Total	2,087,409	2,314,094	2,524,096	2,730,503	2,941,312	3,173,614	3,412,457
K	Irrigation	620,930	589,705	567,272	545,634	524,809	504,695	468,763
	Livestock	13,395	13,395	13,395	13,395	13,395	13,395	13,395
	Manufacturing	28,887	38,162	44,916	56,233	69,264	77,374	85,698
	Mining	23,945	30,620	31,252	31,613	26,964	27,304	27,598
	Municipal	213,303	252,637	304,735	352,737	394,101	439,049	484,170
	Steam Electric	103,875	153,522	156,894	194,396	208,982	214,783	222,058
	Total	1,004,335	1,078,041	1,118,464	1,194,008	1,237,515	1,276,600	1,301,682
L	Irrigation	383,332	379,026	361,187	344,777	329,395	315,143	301,679
	Livestock	25,660	25,954	25,954	25,954	25,954	25,954	25,954
	Manufacturing	100,195	119,310	132,836	144,801	156,692	167,182	179,715

Table 4-7. (Continued)

Projected Water Demands (Acre-feet)								
Region	Category	2000	2010	2020	2030	2040	2050	2060
	Mining	11,757	14,524	15,704	16,454	17,212	17,977	18,644
	Municipal	340,030	395,996	451,111	503,375	547,136	592,343	637,235
	Steam Electric	35,379	50,427	56,792	66,397	78,104	92,378	109,776
	Total	896,353	985,237	1,043,584	1,101,758	1,154,493	1,210,977	1,273,003
N	Irrigation	21,971	20,072	18,611	17,077	15,703	14,470	13,365
	Livestock	8,838	8,838	8,838	8,838	8,838	8,838	8,838
	Manufacturing	54,481	63,820	69,255	73,861	78,371	82,283	88,122
	Mining	11,897	15,150	16,524	16,640	17,490	18,347	19,114
	Municipal	99,951	111,495	122,861	132,063	139,425	146,036	151,474
	Steam Electric	8,799	7,316	14,312	16,733	19,683	23,280	27,664
	Total	205,937	226,691	250,401	265,212	279,510	293,254	308,577

Adapted from TWDB, 2007.

Table 4-8. Projected Shortfall in Water Supply under Drought Conditions (units are in acre-feet per year)

Region/Year	2010	2060
H	279,996	1,119,307
K	246,055	557,311
L	156,598	416,859
N	3,404	53,432

Adapted from TWDB, 2007.

Table 4-9. Costs of Proposed Projects to Mitigate Drought Related Shortfall

Region	Additional Water (acre-feet)	Capital Costs ($)	$/Acre-foot
H	1,300,639	5,460,520,392	4198.34
K	861,930	358,174,068	415.55
L	732,779	5,222,408,000	7126.85
N	149,496	789,515,000	5281.18

Adapted from TWDB, 2007.

4.4. Freshwater Needs for Thermoelectric Power Plants and Carbon Capture Technology

Water consumption for new and existing power plants depends on the type of process used to generate electricity. The Department of Energy (DOE) and the NETL developed several relative water use metrics for power plants as illustrated in Figure 4-13. The types of power plant technologies outlined by DOE/NETL (2007a) include new pulverized coal (PC) and integrated gasification combined cycle (IGCC) plants with and without carbon sequestration technology installed.

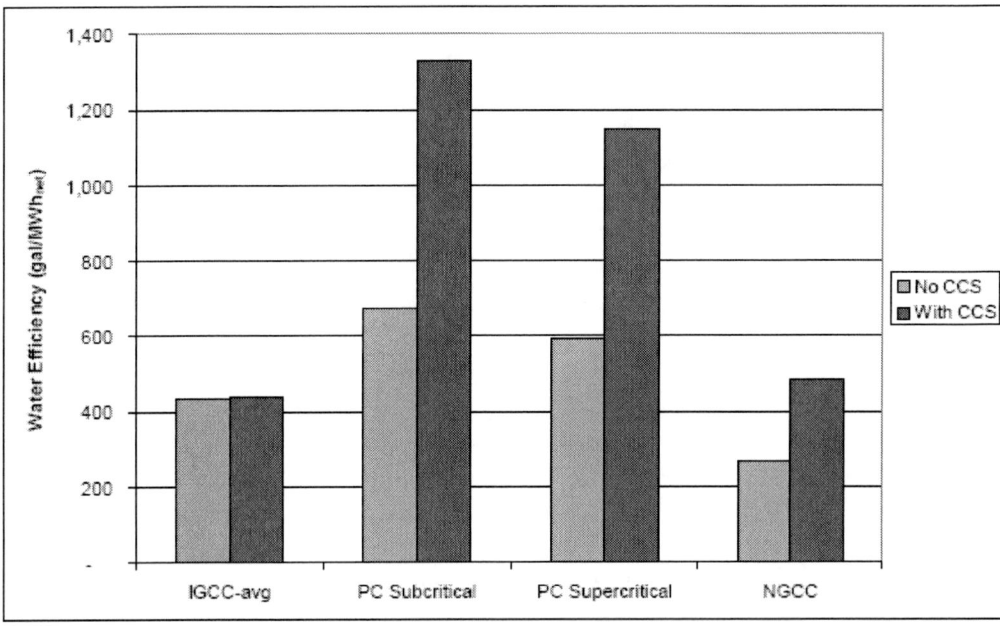

Carbon Capture and Sequestration (CCS); Integrated Gasification Combined Cycle (IGCC); Pulverized Coal (PC); Natural Gas Combined Cycle (NGCC)) (adapted from DOE/NETL, 2007a.

Figure 4-13. Relative Water Usage of New PC and IGCC Plants.

Expansion at power plants will require additional water if wet cooling towers are to be used. Non-freshwater (e.g., brackish water or produced water) will require desalination prior to use. Dry cooling towers could be employed, but there is an energy penalty as well as site-specific operational constraints (DOE, 2007b, c; EPRI, 1982, 2003, 2006).

Another potential solution is to identify a technology that utilizes the waste heat currently sent to the condenser system or in the flue gas for desalination or other process use. This could allow for an alternative water source high in TDS to be utilized economically, as well as help in the energy penalty associated with carbon sequestration. The National Energy Technology Laboratory has been evaluating thermal membrane distillation techniques that utilize power plant waste heat as a desalination technology.

4.5. Detailed Summary of Desalination Options Studies for the Model

Once the potential coupled-use formations within the SECARB region were identified in Table 4-5, feasible desalination options were studied using a basic spreadsheet analysis. All options utilized reverse osmosis (RO) for desalination and varied in the mode of waste water disposal. Currently, RO is the most cost-effective method for treating saline water. Other membrane options and thermal distillation methods exist, but RO has proven to be the most effective option for the price. Thermal membrane distillation is a promising option that could be cost competitive with RO if the energy for the process comes in the form of relatively low-grade power plant waste heat. This technology is currently under evaluation in various forms to determine its feasibility to provide another option for nontraditional water treatment. For this project, the practical options studied were as follows:

Option A	Desalination & Gathering Equipment only; no concentrate disposal
Option B	Desalination & Gathering Equipment only; 59.5 acre evaporation ponds for concentrate disposal
Option C	Desalination & Gathering Equipment only; 3000 ft injection pipeline and well for concentrate disposal
Option D	Desalination & Gathering Equipment – HERO™ (HERO™ +Brine concentrator retrofit (original DiFilippo (2006) report numbers are based on higher TDS)

Each of the options studied required calculations for the capital and operational and maintenance (O&M) costs. Most of the calculations are based on the *USBR Desalting Handbook for Planners* (USBR, 2003). The important parameters for estimating purposes are summarized in Table 4-10. Basically, each treatment scenario assumes a design flow rate of 2.0 MGD, an annual flow rate of 400 Mgal/year based on an 85% plant capacity utilization rate. Additional assumptions regarding capital and O&M costs are included in Table 4-10. The primary difference in costing of the different options comes from the parameters that are site specific, such as groundwater TDS concentration, the distance from the wellhead to the water treatment facility, and the well depth. This working methodology draws heavily from work that first appeared in Kobos et al., 2009, and projects underlying that report.

Summary costs of the four treatment options for each power plant/wellhead choice are shown in the Tables 4-11, 4-12, 4-13, and 4-14. The costs include capital and O&M for each option. It should be noted that concentrate disposal only considers engineering process costs. Issues such as permitting and/or land availability can significantly affect costs or limit options; however, they are beyond the scope of this study. Table 4-15 roles up the significant cost information from Tables 4-11 through 4-14. For this analysis, Option A is considered the baseline case. It includes standard RO treatment with no disposal of the waste concentrate. Option B includes RO treatment with disposal to evaporation ponds with the assumption that there is land available for the ponds. Option C includes RO treatment with concentrate re-injection into the formation, and finally, Option D is the High Efficiency Reverse Osmosis HERO™ in association with a brine concentrator.[4] The brine concentrator eliminates the need for concentrate management at the plant design flow rate of 1850 gpm. If the flow rate were to exceed 4,640 gpm, then evaporation ponds or re-injection would become necessary. Table 4-15 compares Options B, C, and D to the baseline A. It is anticipated that concentrate management will always be required; however, costing without it is included to shed light on concentrate disposal costs. In general, when concentrate management is considered, Option C (RO treatment with concentrate reinjection) is the least expensive option, followed by Option D (the HERO™ System) followed by Option B (RO treatment with evaporation ponds).

Note that the costs vary considerable (as much as 50%) among the options; however, costs vary only slightly across the different sites. This means that RO water treatment costs do not vary significantly when TDS levels are within the desired range for the technology. RO is a proven, well-documented technology with known limitations. Operation outside the prescribed range will result in much higher costs. The cost variation from site to site is a result of variations in TDS levels, well depth, and distance from treatment facility.

Note also that the costs have been developed by using TDS level; however, if an actual system were installed, a more detailed analysis of the water chemistry would be required prior to final design. For example, if the water had high iron or sulfate content, a pre-treatment would likely be required, resulting in increased costs. In conclusion, the costs are representative of typical saline water treatment costs, but a final design must include a more detailed evaluation of input parameters.

These results are an initial analysis for the named power plants, while accounting for a 2.0 MGD desalination facility on site and compares each of the options considered. The desalination costs are in line with both the DiFilippo (2006) study, as well as others (Miller, 2003; USBR, 2003) who have studied the cost of desalination.

Table 4-10. Summary of RO Input Parameters and Cost Components

RO Input Parameter	Description	Value
Water TDS (ppm)	Specific to each well	Actual value
Design flow rate (gpm)	based on 24/7 pumping of brackish aquifer, and % CO_2 capture	1,807
Design flow rate (MGD)	Used 2.0 MGD desalination output (treated water)	
Design annual flow (Mgal/yr)	based on 0.85 plant capacity factor (USBR recommendation)	400
Electrical cost ($/kwh)		0.1
Pipeline distance from brackish well to desal plant (mi)	based on radial distance,	Actual value
Well Depth (ft)	based on well data	Actual value
Cost Components	Reference Source	
Capital Costs:		
Pump & Pipe – produced water gathering capital	Used USBR *Desalting Handbook*, 2009: Figure 9-18	$2,000/ft
Piping from gathering station to desal Plant	Used USBR *Desalting Handbook*, Figure 9-11	$126,810/mi
Concentrate disposal pipeline & well	Used USBR *Desalting Handbook*, Figure 9-11 & 9-13	
Evaporation ponds	Used USBR *Desalting Handbook*, Figure 9-12	
Desalination Total Construction Cost	Used USBR *Desalting Handbook*, Figure 9-7	$2000/ppm TDS for options A-C DiFilippo (2006) value for option D
O&M		
Labor (for 2 MGD)	Used USBR *Desalting Handbook*, Figure 9-37	
Electrical-brackish water reverse osmosis (BWRO) (for 6,000 ppm TDS, 2 MGD)	Used USBR *Desalting Handbook*, Figures 7-8 & 9-45	
Electrical-GW (groundwater) pumping (for 1807 gpm/2 MGD)	Used equations to estimate pump power.	
Membrane replacement	Used USBR *Desalting Handbook*, BWRO	$0.08/1000 gal plant capacity
Chemicals (used surface water)	Used USBR *Desalting Handbook*, Figure 9-41	
Other maintenance	Used USBR *Desalting Handbook*	1.5% of capital

Table 4-11. OPTION A – RO Treatment with No Disposal

				Power Plants			
			Fayette	**Fayette**			
	Coleto	**Coleto**	**Power**	**Power**		**Sam**	**Sam**
Total Cost of Treatment	**Creek**	**Creek**	**Project**	**Project**	**TH**	**Rayburn**	**Rayburn**
(incl. concentrate disposal, Surface & GW	**(deep)**	**(shallow)**	**(deep)**	**(shallow)**	**Wharton**	**(shallow)**	**(deep)**
pumping,	**$/1000**	**$/1000**	**$/1000**	**$/1000**	**$/1000**	**$/1000**	**$/1000**
$U.S. 2000)	**gal**	**gal**	**gal**	**gal**	**gal**	**gal**	**gal**
Annualized Total Capital	$2.44	$2.27	$2.39	$2.28	$1.79	$2.16	$2.80
Annual O&M	$3.35	$3.09	$3.53	$2.90	$2.79	$2.71	$3.33
Electrical	$1.58	$1.35	$1.77	$1.16	$1.14	$0.98	$1.49
Membrane Replacement	$0.08	$0.08	$0.08	$0.08	$0.08	$0.08	$0.08
Chemicals	$0.91	$0.91	$0.91	$0.91	$0.91	$0.91	$0.91
Other	$0.78	$0.75	$0.77	$0.75	$0.66	$0.73	$0.85
Total Cost (O&M+cap)	**$5.79**	**$5.36**	**$5.92**	**$5.19**	**$4.58**	**$4.86**	**$6.13**
			Fayette	**Fayette**			
Cost of Desalination only –	**Coleto**	**Coleto**	**Power**	**Power**	**TH**	**Sam**	**Sam**
includes only equipment &	**Creek**	**Creek**	**Project**	**Project**	**Wharton**	**Rayburn**	**Rayburn**
O&M for desalination (i.e., no ponds, no GW pumping)	**$/1000 gal**	**$/1000 gal**	**$/1000 gal**	**$/1000 gal**	**$/1000 gal**	**$/1000 gal**	**$/1000 gal**
Annualized Total Capital	$1.49	$1.49	$1.49	$1.49	$1.49	$1.49	$1.49
Annual O&M	$2.25	$2.22	$2.38	$1.98	$2.00	$1.88	$2.10
Electrical	$0.48	$0.48	$0.62	$0.24	$0.35	$0.16	$0.26
Membrane Replacement	$0.08	$0.08	$0.08	$0.08	$0.08	$0.08	$0.08
Chemicals	$0.91	$0.91	$0.91	$0.91	$0.91	$0.91	$0.91
Other	$0.78	$0.75	$0.77	$0.75	$0.66	$0.73	$0.85
Total Cost (O&M+cap)	**$3.74**	**$3.71**	**$3.87**	**$3.47**	**$3.49**	**$3.37**	**$3.59**

Table 4-12. OPTION B – RO Treatment with Evaporation Pond Disposal

Total Cost of Treatment (includes concentrate disposal, GW pumping, $U.S. 2000)				Power Plants			
	Coleto Creek (deep)	Coleto Creek (shallow)	Fayette Power Project (deep)	Fayette Power Project (shallow)	TH Wharton	Sam Rayburn (shallow)	Sam Rayburn (deep)
Annualized Total Capital	$4.77	$4.61	$4.72	$4.62	$4.13	$4.49	$5.14
Annual O&M	$3.79	$3.53	$3.97	$3.34	$3.23	$3.14	$3.77
Electrical	$1.58	$1.35	$1.77	$1.16	$1.14	$0.98	$1.49
Membrane Replacement	$0.08	$0.08	$0.08	$0.08	$0.08	$0.08	$0.08
Chemicals	$0.91	$0.91	$0.91	$0.91	$0.91	$0.91	$0.91
Other	$1.22	$1.19	$1.21	$1.19	$1.10	$1.17	$1.29
Total Cost (O&M+cap)	$8.56	$8.14	$8.69	$7.96	$7.36	$7.64	$8.90
			Fayette	Fayette			
	Coleto	Coleto	Power	Power		Sam	Sam
Cost of Desalination only –	Creek	Creek	Project	Project	TH	Rayburn	Rayburn
includes only equipment &	(deep)	(shallow)	(deep)	(shallow)	Wharton	(shallow)	(deep)
O&M for desalination (i.e., no ponds, no GW pumping)							
	$/1000	$/1000	$/1000	$/1000	$/1000	$/1000	$/1000
	gal	gal	gal	gal	gal	gal	gal
Annualized Total Capital	$1.49	$1.49	$1.49	$1.49	$1.49	$1.49	$1.49
Annual O&M	$2.68	$2.65	$2.82	$2.42	$2.44	$2.32	$2.54
Electrical	$0.48	$0.48	$0.62	$0.24	$0.35	$0.16	$0.26
Membrane Replacement	$0.08	$0.08	$0.08	$0.08	$0.08	$0.08	$0.08
Chemicals	$0.91	$0.91	$0.91	$0.91	$0.91	$0.91	$0.91
Other	$1.22	$1.19	$1.21	$1.19	$1.10	$1.17	$1.29
Total Cost (O&M+cap)	$4.17	$4.14	$4.31	$3.91	$3.93	$3.81	$4.03

Table 4-13. OPTION C – RO treatment with Concentrate Re-injection

				Power Plants			
			Fayette	Fayette			
	Coleto	Coleto	Power	Power		Sam	Sam
Total Cost of Treatment	**Creek**	**Creek**	**Project**	**Project**	**TH**	**Rayburn**	**Rayburn**
(incl. concentrate disposal,	**(deep)**	**(shallow)**	**(deep)**	**(shallow)**	**Wharton**	**(shallow)**	**(deep)**
GW pumping, $U.S. 2000)	**$/1000 gal**	**$/1000 gal**	**$/1000 gal**	**$/1000 gal**	**$/1000 gal**	**$/1000 gal**	**$/1000 gal**
Annualized Total Capital	$2.88	$2.21	$2.83	$2.73	$2.23	$2.60	$3.25
Annual O&M	$3.44	$2.91	$3.62	$2.99	$2.88	$2.79	$3.42
Electrical	$1.58	$1.35	$1.77	$1.16	$1.14	$0.98	$1.49
Membrane Replacement	$0.08	$0.08	$0.08	$0.08	$0.08	$0.08	$0.08
Chemicals	$0.91	$0.91	$0.91	$0.91	$0.91	$0.91	$0.91
Other	$0.87	$0.57	$0.86	$0.84	$0.75	$0.82	$0.94
Total Cost (O&M+cap)	**$6.32**	**$5.12**	**$6.45**	**$5.72**	**$5.11**	**$5.39**	**$6.66**
			Fayette	Fayette			
	Coleto	Coleto	Power	Power		Sam	Sam
Cost of Desalination only –	**Creek**	**Creek**	**Project**	**Project**	**TH**	**Rayburn**	**Rayburn**
includes only equipment &	**(deep)**	**(shallow)**	**(deep)**	**(shallow)**	**Wharton**	**(shallow)**	**(deep)**
O&M for desalination (i.e., no ponds, no GW pumping)	**$/1000 gal**	**$/1000 gal**	**$/1000 gal**	**$/1000 gal**	**$/1000 gal**	**$/1000 gal**	**$/1000 gal**
Annualized Total Capital	$1.49	$1.49	$1.49	$1.49	$1.49	$1.49	$1.49
Annual O&M	$2.34	$2.04	$2.47	$2.07	$2.09	$1.97	$2.19
Electrical	$0.48	$0.48	$0.62	$0.24	$0.35	$0.16	$0.26
Membrane Replacement	$0.08	$0.08	$0.08	$0.08	$0.08	$0.08	$0.08
Chemicals	$0.91	$0.91	$0.91	$0.91	$0.91	$0.91	$0.91
Other	$0.87	$0.57	$0.86	$0.84	$0.75	$0.82	$0.94
Total Cost (O&M+cap)	**$3.82**	**$3.53**	**$3.96**	**$3.56**	**$3.58**	**$3.46**	**$3.68**

Table 4-14. OPTION D – HERO™ System

				Power Plants			
			Fayette	Fayette			
	Coleto	Coleto	Power	Power		Sam	Sam
Total Cost of Treatment	**Creek**	**Creek**	**Project**	**Project**	**TH**	**Rayburn**	**Rayburn**
(incl. concentrate disposal,	**(deep)**	**(shallow)**	**(deep)**	**(shallow)**	**Wharton**	**(shallow)**	**(deep)**
GW pumping, $U.S. 2000)	**$/1000 gal**	**$/1000 gal**	**$/1000 gal**	**$/1000 gal**	**$/1000 gal**	**$/1000 gal**	**$/1000 gal**
Annualized Total Capital	$2.82	$2.66	$2.77	$2.67	$2.18	$2.54	$3.19
Annual O&M	$4.48	$4.22	$4.66	$4.03	$3.92	$3.83	$4.46
Electrical	$2.62	$2.39	$2.81	$2.20	$2.18	$2.03	$2.53
Membrane Replacement	$0.08	$0.08	$0.08	$0.08	$0.08	$0.08	$0.08
Chemicals	$0.93	$0.93	$0.93	$0.93	$0.93	$0.93	$0.93
Other	$0.85	$0.82	$0.84	$0.82	$0.73	$0.80	$0.92
Total Cost (O&M+cap)	**$7.30**	**$6.88**	**$7.43**	**$6.70**	**$6.10**	**$6.37**	**$7.64**

Table 4-14. (Continued)

Cost of Desalination only – includes only equipment & O&M for desalination (i.e., no ponds, no GW pumping)	Coleto Creek (deep) $/1000 gal	Coleto Creek (shallow) $/1000 Gal	Fayette Power Project (deep) $/1000 gal	Fayette Power Project (shallow) $/1000 gal	TH Wharton $/1000 gal	Sam Rayburn (shallow) $/1000 gal	Sam Rayburn (deep) $/1000 gal
Annualized Total Capital	$1.87	$1.87	$1.87	$1.87	$1.87	$1.87	$1.87
Annual O&M	$3.38	$3.35	$3.51	$3.11	$3.13	$3.01	$3.23
Electrical	$1.52	$1.52	$1.66	$1.28	$1.39	$1.20	$1.31
Membrane Replacement	$0.08	$0.08	$0.08	$0.08	$0.08	$0.08	$0.08
Chemicals	$0.93	$0.93	$0.93	$0.93	$0.93	$0.93	$0.93
Other	$0.85	$0.82	$0.84	$0.82	$0.73	$0.80	$0.92
Total Cost (O&M+cap)	**$5.25**	**$5.22**	**$5.39**	**$4.98**	**$5.00**	**$4.88**	**$5.11**

Table 4-15. Summary of Total Desalination Costs by Option

	POWER PLANTS						
OPTION (including disposal & GW pumping) ($U.S. 2000)	**Coleto Creek (deep)**	**Coleto Creek (shallow)**	**Fayette Power Project (deep)**	**Fayette Power Project (shallow)**	**TH Wharton**	**Sam Rayburn (shallow)**	**Sam Rayburn (deep)**
	TOTAL ANNUALIZED COSTS (CAP + O&M) ($1000/gal)						
A	$5.79	$5.36	$5.92	$5.19	$4.58	$4.86	$6.13
B	$8.56	$8.14	$8.69	$7.96	$7.36	$7.64	$8.90
C	$6.32	$5.12	$6.45	$5.72	$5.11	$5.39	$6.66
D	$7.30	$6.88	$7.43	$6.70	$6.10	$6.37	$7.64
% Difference vs. Baseline Case (Option A)							
A-		-	-	-	-	-	-
B	48 %	52 %	47 %	53 %	61 %	57 %	45 %
C	9 %	-4 %	9 %	10 %	12 %	11 %	9 %
D	26 %	28 %	26 %	29 %	33 %	31 %	25 %
OPTION (with no disposal or GW pumping)	**TOTAL ANNUALIZED COSTS (CAP + O&M) ($1000/gal)**						
A	$3.74	$3.71	$3.87	$3.47	$3.49	$3.37	$3.59
B	$4.17	$4.14	$4.31	$3.91	$3.93	$3.81	$4.03
C	$3.82	$3.53	$3.96	$3.56	$3.58	$3.46	$3.68
D	$5.25	$5.22	$5.39	$4.98	$5.00	$4.88	$5.11
	% Difference vs. Baseline Case (Option A)						
A-		-	-	-	-	-	-
B	11 %	12 5	11 5	13 %	13 %	13 %	12 %
C	2 %	-5 5	2 5	3 %	3 %	3 %	3 %
D	40 %	41 5	39 5	44 %	43 %	45 %	42 %

CONCLUSION

The analysis determined that using a sequential framework examining the geologic, power plant, and geomodeling criteria, several regions of the U.S. have potential sites for a coupled carbon dioxide sequestration, water extraction, and treatment system. They all have unique attributes and issues requiring a thorough site-specific examination of the geology type, depth, geochemistry, and extent before placing a full-scale coupled-use system as illustrated in Figure 2-1 on these sites. A few of the salient lessons learned from the data collection and analysis from the geological assessment include: (1) more data on deep saline formation water TDS (e.g., geochemistry) may need to be collected to more fully evaluate regional sequestration options for carbon dioxide and (2) the relative, calculated size of the formations (and hence their useful lifetime for sequestering carbon or extracting water for treatment and subsequent use) is highly dependent on assumptions that must be verified (such as formation porosity and permeability) perhaps through pilot projects. With this information, a more accurate assessment could be developed for regions of the southwestern United States with respect to CO_2 storage in saline formations, the potential for water extraction and use, and the interactive effects between these two types of systems as they relate to power plant demands for water today and in the coming decades.

REFERENCES

Allis, R.G., T.C. Chidesy, C. Morgan, J. Moore, S.P. White, 2003, "CO2 Sequestration Potential beneath Large Power Plants in the Colorado Plateau-Southern Rocky Mountain Region, USA," Proceedings Second Annual Conference on Carbon Sequestration, Alexandria VA, May 5–8 (sponsored by the U.S. Department of Energy).

Beckman, J.D., and A.K. Williamson, 1990, *Salt-Dome Locations in the Gulf Coastal Plain South-Central United States*, USGS Water Resources Investigations Report 90-4060, http://pubs.er.usgs.gov/djvu/WRI/wrir_90_4060.djvu As of October 26, 2009.

Department of Energy (DOE), 2003, *Handbook on Coal Bed Methane Produced Water. Management and Beneficial Use Alternatives*, Chapter 2: "Introduction to Coal Bed Methane," Table 2-3, pp. 2-10, Prepared by ALL Consulting, Tulsa Oklahoma.

Department of Energy (DOE), 2006a, *Energy demands on Water Resources, Report to Congress on the Interdependency of Energy and Water*, http://www.sandia.gov/energy-water/congress_report.htm As of October 19, 2009.

Department of Energy (DOE), 2006b, *Estimating Freshwater Needs to Meet Future Thermoelectric Generation Requirements,* DOE/NETL-2006/1235.

Department of Energy / National Energy Technology Laboratory (DOE/NETL), 2006c, *Use of Produced Water in Recirculating Cooling Systems at Power Generating Facilities,* http://www.netl.doe.gov/technologies/coalpower/ewr/water/pp-mgmt/epri.html As of March 2006.

Department of Energy / National Energy Technology Laboratory (DOE/NETL), 2007a, *Cost and Performance Baseline for Fossil Energy Plants. Volume 1. Bituminous Coal and Natural Gas to Electricity Final Report*, DOE/NETL-2007/1281. May 2007, revised August 2007.

Department of Energy / National Energy Technology Laboratory (DOE/NETL), 2007b, *Estimating Freshwater Needs to Meet Future Thermoelectric Generation Requirements. 2007 Update Forecasts*, DOE/NETL-400/2007/1304.

Department of Energy / National Energy Technology Laboratory (DOE/NETL), 2007c, *Power Plant Water Usage and Loss Study*, August 2005, revised May 2007.

Dickey, P.A., 1969, Increasing concentration of subsurface brines with depth, *Chemical Geology* 4, pp. 361-370.

DiFilippo, M., 2006, Semi-Annual Technical Progress Report, October 1, 2005 to March 31, 2006, *Use of Produced Water in Recirculating Cooling Systems at Power Generating Facilities*, NETL, 2006. http://www.netl.doe.gov/technologies/coalpower/ewr/water/pp-mgmt/epri.html As of October 23, 2007.

Dutton, A.R., B. Harden, J.-P. Nicot, D. O'Rourke, S.W. Tinker, 2003, Groundwater availability model for the central part of the Carrizo-Wilcox Aquifer in Texas. Final report, prepared for Texas Water Development Board. http://www.twdb.state.tx.us /Gam/czwx_c/CZWX_c_Part1_97_Finalrev2a.pdf As of October 26, 2009.

Electric Power Research institute (EPRI), 1982, *Design and Operating Manual for Cooling-Water Treatment*, CS-2276, Section 2.

Electric Power Research Institute (EPRI) and the California Energy Commission (CEC), 2003, *Use of Degraded Water Sources as Cooling Water Plants*, October 2003. http:// www.energy.ca.gov/reports/2004-02-23_500-03-110.PDF As of October 26, 2009.

Electric Power Research Institute (EPRI), 2006, *Cooling Tower Water Quality Parameters for Degraded Water*, April 2006, http://www.energy.ca.gov/reports/2004-02-23_500-03-110.PDF.

Environmental Protection Agency (EPA), 2003a, *Study to Evaluate the Impacts to USDWs by Hydraulic Fracturing of Coalbed Methane Reservoirs*, EPA 816-F-04-017. http:// www.epa.gov/OGWDW/uic/wells_coalbedmethanestudy.html As of October 26, 2009.

Environmental Protection Agency (EPA), 2003b, *Draft Programmatic Environmental Impact Statement on Mountaintop Mining/Valley Fills in Appalachia – 2003*. Appendix C. http://www.epa.gov/Region3/mtntop/eis2003.htm As of October 26, 2009.

Environmental Protection Agency (EPA), 2004, *Evaluation of Impacts to Underground Sources of Drinking Water by Hydraulic Fracturing of Coalbed Methane Reservoirs,* Attachment 2: "The Black Warrior Basin") EPA 816-R-04-003. http://www.epa.gov/ogwdw/ uic/pdfs/cbmstudy_attach_uic_attach02_blkwarrior.pdf As of October 19, 2009.

Environmental Protection Agency (EPA), 2008, *EPA Proposes New Requirements for Geologic Sequestration of Carbon Dioxide,* EPA 816-F-08-032.

Environmental Protection Agency (EPA) eGRID database, 2009, Version 1.0 to 1.1. http:// www.epa.gov/cleanenergy/energy-resources/egrid/index.html As of September, 2009.

Geological Survey of Alabama, 2006, Special Map 220A, Tuscaloosa, Alabama. GIS data available at: http://www.gsa.state.al.us

Georgia Geologic Survey, 1999, Digital Geologic Map of Georgia (ver. 2), Documentation Report 99-20. Georgia Department of Natural Resources, Environmental Publication Division. Atlanta, GA.

Hesse, M.A., F.M. Orr, Jr., and H.A. Tchelepi, 2008, Gravity currents with residual trapping, *J. Fluid Mechanics* 611, pp. 35-60.

Hesse, M.A., F.M. Orr, Jr., and H.A. Tchelepi, 2009, Gravity currents with residual trapping, *Energy Procedia* 1, pp. 3275-3281.

Hovorka, S. D., 1999, *Optimal Geological Environments for Carbon Dioxide Disposal in Saline Aquifers in the United States*, Draft final Contractor Report (DOE Contract Number AC26-98FT40417), available: OSTI as DE00834155. http://www.osti.gov/energycitations/product.biblio.jsp?osti_id=834155 As of October 26, 2009.

Hovorka, S. D., M.L. Romero, A.G. Warne, W.A. Ambrose, T.A. Tremblay, R.H. Treviño, and D. Sasson, 2000, *Sequestration of Greenhouse Gases in Brine Formations,* Texas Bureau of Economic Geology http://www.beg.utexas.edu/environqlty/co2seq/ As of October 26, 2009.

Kelley, V.A., N.E. Deeds, D.G. Fryar, J.-P. Nicotl, T.L. Jones, A.R. Dutton G. Bruehl, T. Unger-Holtz, and J. L. Machin, 2004, *Groundwater Availability Models for the Queen City and Sparta Aquifers*: final report prepared for the Texas Water Development Board by INTERA Inc.

Kobos, P.H., J.L. Krumhansl, T.A. Dewers, J.E. Heath, M.A. Cappelle, D.J. Borns, G.T. Klise, B.P. Dwyer, and A. McNemar, 2009, "Combining Power Plant Water Needs and Carbon Storage using Saline Formation: An Assessment Tool," 8th Annual Conference on Carbon Capture and Sequestration, DOE/NETL, SAND2009-2557C, Pittsburgh, PA, May 4−7.

Mancini, E., and D. Goddard, D., 2007, *Basin Analysis and Petroleum System Characterization and Modeling, Interior Salt Basins, Central and Eastern Gulf of Mexico*, University of Alabama and Louisiana State University, PCCT Network News, July 2007 http://www.pttc.org/newsletter/2qtr2007/v13n2p9.htm As of October 26, 2009.

Meyer, F.W., 1989, *Hydrogeology, Ground-Water Movement, and Subsurface Storage in the Florida Aquifer System in Southern Florida*, USGS Prof. Paper 1403-G: http://sofia.usgs.gov/publications/papers/pp1403g/1403-G.pdf As of October 26, 2009.

Milici, R.C., 2004, *Assessment of Appalachian Basin Oil and Gas Resources: Carboniferous Coal-bed Gas Total Petroleum System*, U.S. Geological Survey Open-File Report 2004-1272 http://pubs.usgs.gov/of/2004/1272/2004-1272.pdf As of October 26, 2009.

Miller, J., 2003, *Review of Water Resources and Desalination Technologies,* Sandia National Laboratories, SAND2003-0800.

Murrie, G.W., W.P. Diamond, and S.W. Lambert, 1976, *Geology of the Mary Lee Group of Coalbeds, Black Warrior Coal Basin, Alabama, Report of Investigations 8189*, Bureau of Mines Report, Pittsburgh Mining and Safety Research Center, Pittsburgh, PA, Fig. 2: http://www.cdc.gov/niosh/mining/pubs/pdfs/ri8189.pdf As of October 19, 2009.

National Carbon Sequestration Atlas (NATCARB) data files, 2008, National Energy Technology Laboratory. Regional Partnership Information and GIS Data, http://www.natcarb.org As of October 19, 2009.

National Energy Technology Laboratory (NETL), 2008a, *Carbon Sequestration Atlas of the United States and Canada,* http://www.netl.doe.gov/technologies/carbon_seq/refshelf/atlas/ATLAS.pdf As of October 19, 2009.

National Energy Technology Laboratory (NETL), 2008b, Southeast Regional Carbon Sequestration Partnership—Validation Phase: "Project Facts" Updated 04/2008, http://www.netl.doe.gov/publications/factsheets/project/Proj442.pdf As of October 19, 2009.

Navoy, A.S., 1986, *Hydrogeologic Data from a 2000-Foot Deep Core Hole at Polk City, Green Swamp Area, Central Florida*, USGS Water-Resources Investigations Report 84-4257.

Nemeth, K.J., 2005, Southeast Regional Carbon Sequestration Partnership (SECARB), *Semi-Annual Report, Start Date: 10/01/04, End Date: 03/31/05,* DOE Award Number: DE-FC26-03NT41980, Published April, 2005, http://www.osti.gov/bridge/servlets/purl /840356-JVeYkl/native/840356.pdf

Nicot, J.-P., 2008, Evaluation of large-scale CO2 storage on fresh-water sections of aquifers: An example from the Texas Gulf Coast Basin, *International Journal of Greenhouse Gas Control* 2, pp. 582-593.

Nicot, J.-P., S.D. Hovorka and J.-W. Choi, 2009, Investigation of water displacement following large CO2 sequestration operations. *Energy Procedia* 1, pp. 4411-4418.

O'Neil, P.E., H.R. Isaacson, and J.M. Evans, J.M., 1992, "Surface Discharge of Coalbed Methane Produced Waters in the Warrior Basin of Alabama, the Cedar Cove Model," in *Proceedings of the 1992 International Produced Water Symposium*, held February 4-7, 1992, in San Diego, California, pp. 329-342 (Table 1, P. 332), http://books.google.com/ books?id=0BFDzWFL8t8C&pg=PA340&lpg=PA340&dq=cedar+cove+model+alabama &source=web&ots=Xa5QvIjE1l&sig=bFqw02YfzowWvReyqckaGvtXj-k&hl=en&sa =X&oi=book_result&resnum=1&ct=result#PPA332,M1 As of October 19, 2009.

O'Reilly, A.M., R.M. Spechlerk, and B.E. McGurk, B.E., 2002, *Hydrogeology and Water-Quality Characteristics of the Lower Floridan Aquifer in East-Central Florida*, USGS Water-Resources Investigations Report 02-4193, 60 pp.

Ortiz, I., T.R. Weller, R.V. Anthony, J. Frank, J., and D. Nakles1993, "Disposal of Produced Waters: Underground Injection Option in the Black Warrior Basin," in *Proceedings 1993 Coalbed Methane Symposium*, University of Alabama (Tuscaloosa), pp. 339-364.

Pashin, J.C., W.E. Ward II, R.B. Winston, R.V. Chandler, D.E. Bolin, K.E. Richter, W.E. Osborne, and J.C. Sarnecki, 1991, Regional analysis of the Black Creek-Cobb coalbed-methane target interval, Black Warrior basin, Alabama, *AL Geol. Surv.,* Bull. 145, pp. 127.

Pashin, J.C., and F. Hinkle, F, 1997, *Coalbed Methane in Alabama,* Geological Survey of Alabama Circular 192, pp. 71.

Pashin, J.C., R.E. Carroll, R.H. Groshong, Jr., D.E. Raymond, M. McIntyre, and J.W. Payton, 2003, *Geologic Screening Criteria For Sequestration of CO2 in Coal: Quantifying Potential of the Black Warrior Coalbed Methane Fairway, Alabama, Geologic Survey of Alabama*, Final Report, U.S. Department of Energy, National Energy Technology Laboratory contract DE-FC26-00NT40927, http://www.ogb.state.al.us/CO2/CO2page /40927R03.pdf As of October 19, 2009.

Pashin, J.C., and J.W. Payton, 2005, *Geological sinks for carbon sequestration in Alabama, Mississippi, and the Florida panhandle: Alabama Geological Survey*, Final Report to Southern States Energy Board, Subgrant SSEB-NT41980-997-GSA-2004-00. GIS data available at: http://portal.gsa.state.al.us/CO2/SECARB1/secarb1_files/Phase_I_GIS.zip October 19, 2009.

Pashin, J.C., 2007, "Hydrodynamics of coalbed methane reservoirs in the Black Warrior Basin: Key to understanding reservoir performance and environmental issues," *Applied Geochemistry*, Volume 22, Issue 10, October 2007, pp. 2257-2272.

Pearson, F.J., Jr., and D.E. White, 1967, *Carbon 14 Ages and Flow Rates of Water in Carrizo Sand, Atascosa County, Texas,* Water Resources Research, v. 3 (1), pp. 251-261, http://www.agu.org/pubs/crossref/1967/WR003i001p00251.shtml October 19, 2009.

Pruess, K., 2005, *ECO2N: A TOUGH2 fluid property module for mixtures of water, NaCl, and CO2*, Lawrence Berkeley National Laboratory, LBNL-43134, 66 pp.

Pruess, K., C. Oldenburg, and G. Moridis, 1999, *TOUGH2 User's Guide*, Version 2. Lawrence Berkeley National Laboratory, LBNL-43134, 198 pp.

Smyth, R.C., S.D. Hovorka, T. Meckel, C. Breton, J.G. Paine, G.R. Hill, J.R. Andrews, S. Lakshminarasimhan, H. Herzog, H. Zhang, W. Li, A. Kalinowski, P. Thayer, and T. Temples, 2007, *Potential Sinks for Geologic Storage of Carbon Dioxide Generated in the Carolinas*, Prepared for the Southern States Energy Board and Electric Power Research Institute, http://www.beg.utexas.edu/environqlty/co2seq/pubs_presentations/Carolinas Summary_16April07.pdf As of October 19, 2009.

Southeast Regional Carbon Sequestration Partnership (SECARB), 2008, SECARB Phase III Tuscaloosa Formation CO2 Storage Project: Factsheet for Partnership field validation test. http://www.netl.doe.gov/publications/proceedings/07/rcsp/factsheets/3- SECARB_ Large%20Scale%20Saline%20Formation%20Demo.pdf As of October 19, 2009.

Spechler, R.M., and S.E. Kroening, 2006, *Hydrology of Polk County, Florida*, USGS Scientific Investigations Report 2006-5320, 114 pp. http://pubs.usgs.gov/sir/2006 /5320/pdf/sir2006-5320.pdf As of October 19, 2009.

Sprinkle, C.L., 1989, *Geochemistry of the Floridian Aquifer System in Florida and in Parts of Georgia, South Carolina, and Alabama*, USGS Professional Paper 1403-I, 105 pp. http://sofia.usgs.gov/publications/papers/pp1403i/dissolvedsolids.pdf As of October 19, 2009.

Texas Water Development Board (TWDB) database, 2002, Well Information/Groundwater Data, http://www.twdb.state.tx.us/GwRD/waterwell/well_info.asp As of October 19, 2009.

Texas Water Development Board (TWDB), 2003, *Brackish Groundwater Manual for Texas Regional Water Planning Groups*, by: LBG-GUYTON ASSOCIATES in association with NRS Consulting Engineers; p. 50-56. http://www.twdb.state.tx.us/RWPG/rpgm_rpts /2001483395.pdf As of October 19, 2009.

Texas Water Development Board (TWDB), 2007, *2007 State Water Plan*, available at: http://www.twdb.state.tx.us/wrpi/swp/swp.htm, As of August 25, 2009.

United States Geological Survey (USGS), 1999, *U.S. Geological Survey GROUND WATER ATLAS of the UNITED STATES,* On-line version can be found at: http://pubs.usgs.gov/ha/ha730/gwa.html As of October 19, 2009.

United States Geological Survey (USGS), 2002, *Produced Water Database: Provisional Release*, May 2002, Compiled by George N. Breit; Web authoring by Chris Skinner http://energy.cr.usgs.gov/prov/prodwat/index.htm As of October 19, 2009.

United States Bureau of Reclamation (USBR), 2003, *Desalting Handbook for Planners*, 3rd ed.

Veil, J.A., 2007, *Use of Reclaimed Water for Power Plant Cooling*, ANL/EVS/R-07/3, prepared for the U.S. DOE, National Energy Technology Laboratory, August, 66 pp.

Veil, J., J. Kupar, and M. Puder, 2003, *Use of Mine Pool Water for Power Plant Cooling*, prepared by Argonne National Laboratory for the U.S. Department of Energy, National Energy Technology Laboratory, September, 62 pp.

Veil, J.A., and M.G. Puder, 2006, *Potential Ground Water and Surface Water Impacts from Oil Shale and Tar Sands Energy-Production Operations*, ANL/EVS/R-06/9, prepared for the Ground Water Protection Council, October, 50 pp.

Vernon, R. O., 1970, *The Beneficial Uses of Zones of High Transmissivities in the Florida Subsurface for Water Storage and Waste Disposal,* Florida Geological Survey, Information Circular No. 70, 39.

Wolfram, S., 2003, *The Mathematica Book*, 5th ed., Wolfram Media/Cambridge University Press, 1464 pp.

APPENDIX A

An alternate approach to locating potential study sites in the SECARB area was made possible by gaining the ability to access the entire contents of the NATCARB brine chemistry database. However, the database initially needed to be processed so that just useful entries remained. This involved deleting duplicate entries (based on "ID_Code"), entries that lacked locations (no latitude or longitude), entries that lacked well-depth information, and entries with insufficient water chemistry data. After this, three approaches were used to extract useful insights from the database.

Initially, all data pertinent to the SECARB region were sorted by salinity level into three groupings: data reflecting salinities less than 10,000 parts per million dissolved salts (protected by regulation as potential drinking water resources), data reflecting salinities between 10,000 and 35,000 ppm TDS (suitable to programmatic uses), and data with TDS values greater than 35 ppb TDS (too salty to be desalinated economically). Data in each grouping were then "mapped," plotted based on their longitude and latitude, to see if different geographic regions could be defined where one might encounter high- or low-salinity waters (Figure A1 a,b,c). However, the outcome of this assessment was that the locations for the three classes of fluids overlapped and that geographic areas with the sorts of brackish waters useful to this program could not be uniquely identified with this strategy.

The second approach involved trying to find discriminating locations for different brine classes based on the depth of the wells; for example, shallower wells might be presumed to be less saline. From Florida west to the Mexican border, the layers of porous sedimentary rocks tilt gently downward toward the south. Thus, inland the formations change from old Cretaceous- age strata on the surface while these same rocks are buried many thousands of feet deep beneath Tertiary sediments on the Gulf Coast. The process of assessing vertical salinity gradients then consisted of making plots of depth vs. distance northward from the southern boundary of the cluster of well locations in question (Figure A1 a,b,c) for each of the salinity classes identified earlier. For the wells located in west Texas, the plotted "latitude" (e.g., the north-south distance used on the horizontal axis of the plot) first had to be corrected for the fact that the pattern of well locations bends to the south as one works westward across the state.

Figure A2 a,b,c (south-central Louisiana) and Figure A3 a,b,c (south Texas) are typical of the results from such investigations. Basically, as with the map views, this approach also failed to reveal any significant discrimination (either vertically or laterally) between wells having low, intermediate, and high salinities. The one caveat is that for the band of wells closest to the Gulf Coast, the ratio of low- to high-salinity wells seems slightly higher in the northernmost sections of the plots. This agrees with the earlier identified potential for a reservoir of brackish underground water in the Carrizo-Wilcox Formation north of the well locations compiled in the NATCARB database (e.g., in the northern part of the yellow box outlined in Figure A1 a,b,c). The brackish water-containing formations identified earlier in the report would not have been expected to show up in a NATCARB-based analysis (NATCARB, 2008) because NATCARB (at least in this region) contains information on produced waters (e.g., waters co-produced during hydrocarbon—gas or oil—production), and no hydrocarbons are produced from the CarrizoWilcox in the area identified in the main body of the report.

The final approach to locating areas with brackish waters at sufficient depth returned to the concept of making "maps" (e.g., plotting longitude vs. latitude of wells), but this time, the database was presorted so that wells would be grouped by the formation in which they were completed. In doing so, several features became apparent. First, not all the wells that passed the initial sort criteria actually have useable formation names or formation ages entered in the NATCARB database. In fact, only about 70% of the wells have useable formation names associated with them, and only about 64% of the wells have a geologic age for the rock they were completed in. Further statistical analysis is made difficult because occasionally a formation name will be provided without its corresponding age and vice versa.

Second, in going through the resulting tabulation (ultimately to locate formations with an auspiciously large number of lower salinity hits), it became apparent that only a few formations contained data distributed over a large enough area to define the sorts of regional hydrologic trends useful to this program. In fact, the NATCARB database contains numerous clusters of wells with virtually identical longitudes and latitudes but without stratigraphic correlations extending beyond the boundary of a particular oil or gas field. Of the formations that appeared on a regional basis, only two (the Tertiary-age Claiborne Formation, Figure A4, and the Cretaceous-age Vicksburg Formation, Figure A5) seemed to contain areas where one might expect to find *relatively* uncontaminated low- to intermediate-salinity waters at appropriate depths. All the other (far more numerous assessments) followed the typical pattern where, low-, intermediate-, and high-TDS wells are thoroughly intermixed. Additionally, even where favorable areas were defined (Figures A4 and A5), the designation is generally based on fewer than ten wells (or well clusters) spread out for more than a hundred miles along a relatively narrow trend.

Thirdly, although the regional maps (Figure A1 a,b,c) seemingly contain dense clusters of points, the reality is that individual wells (or dense well clusters with the same assigned longitude and latitude) are most often separated by many miles. On one hand, this makes ruling out the possibility that even a single intermediate TDS well hit could reflect the presence of a *locally* substantial brackish water resource. On the other hand, verifying this may require supplementing the data contained within NATCARB to complete the analysis.

To summarize, relative to applying NATCARB data in the SECARB region:

a) Coverage provided by the NATCARB database is incomplete and provides essentially no coverage for large areas of the SECARB region where neither oil nor gas wells have been drilled.
b) Where pertinent NATCARB data do exist, the dominant picture is that *on a regional scale,* a first-order analysis will likely find that useable wells are intermixed with

both higher- and lower-salinity wells. Consequently, instead of trying to define specific formations as potential sources for brackish water, additional analyses may prove more useful by attaching probabilities to tracts of land that define the likelihood of successfully drilling into a body of brackish water.

c) Difficulties in geographical discrimination notwithstanding, the NATCARB database in total does suggest that brackish water resources in the SECARB region are larger, and more widely distributed, than had been deduced previously based on our review of the conventional geo-hydrologic literature.
d) d. Because most wells are significant distances from the next well listed in the NATCARB database, even a single well with reliable, detailed information might still identify a locally important brackish water resource. Hence, an alternate search strategy for future analyses might be to find the nearest well with an acceptable TDS and assume a storage capacity defined by a circle with a radius of half the distance to the nearest well that is not suitable. Additionally, the distance to the nearest well in the same formation that is unsuitable could be examined.

Table A1. Wilcox & Claiborne Formations Sorted by TDS 10,000 – 35,000 ppm TDS and Well Depth

Record # "UNIQID"	Upper Depth	Lower Depth	TDS	pH	HCO3 ppm	SO4 ppm	Cl ppm	Na ppm	Ca ppm	Mg ppm
---	*9869*	*9928*	*20.8*	*7.2*	*1066*	*507*	*11310*	*7304*	*475*	*68*
42019052	7800		33.2	7	672	98	19560	12379	480	31
42251942	8184	8190	33.2	5.55	0	0	20402	10595	1547	403
42016434	9234	9282	33.1	6.8	1248	49	19020	12600	170	22
42006708	10261	10270	32.7	7	798	74	18845	12042	374	39
42012431	10240	10278	32.6	6.7	670	8	17665	136	10054	20
42257510	8802	8815	31.7	7.4	783	42	18681	11258	661	262
42257780	10680		30.7	5.9	163	214	18350	11063	690	56
42019111	9440	9452	29.9	6.8	1230	0	17100	11300	240	0
42018947	9440	9452	29.6	6.8	1190	0	17100	11300	180	12
42016677	8578	8588	28.3	8.7	1111	36	16260	10614	170	101
42009946	8400		27.4	7.62	1681	50	15170	10232	128	10
42017147	7676	7680	26.5	7	1415	-3	15000	9562	400	122
42006717	8379	8382	26.5	8.2	1733	55	14591	9619	119	200
42006714	8379	8382	26.4	7.6	1728	23	14544	9933	96	22
42006716	8379	8382	26.2	7.3	1693	21	14468	9826	129	31
42257778	10620	10680	25.9	7	814	76	15000	9678	201	84
42011895	10261	10270	25.8	7.3	624	26	14659	9239	318	74
42006712	8382	8385	25.7	7.9	1815	33	14435	9821	157	33
42257677	9335	9380	25.7	6.8	913	-3	14950	9376	452	77
42006713	8382	8385	25.6	7.9	1853	36	14182	9698	141	30
42256923	11480	11491	25.6	6.65	216	97	15355	8311	1389	96
42258082	9598	9608	25.4	6.45	242	0	15360	8673	895	160
42257724	8972	8976	24.1	6.85	550	560	13800	8080	934	122
42258089	9100	9125	24.0	7.7	1144	8	13653	8983	170	57
42005127	9503	9520	23.9	7.3	1320	3150	10400	7610	1370	49
42004344	10207		23.7	8	1460	11000	3300	7560	300	36
42006706	8364	8369	23.6	7.5	2074	26	12606	8658	228	27

Record # "UNIQID"	Upper Depth	Lower Depth	TDS	pH	HCO3 ppm	SO4 ppm	Cl ppm	Na ppm	Ca ppm	Mg ppm
42009448	10052	10090	23.0	7.31	1258	8	12949	8462	195	59
42257825	10590	10878	22.9	6.42	285	20	13700	7820	604	190
42257822	10858	10938	22.7	7.3	431	0	13482	8238	356	122
42258080	8656	8665	22.0	7.72	1290	75	12200	8290	121	0
42018945	9448	9453	21.9	6.8	982	375	11600	8570	310	97
42018717	9295	9328	21.9	7.4	1830	114	11700	8110	80	68
42257473	11070	11080	21.7	7.55	1410	320	11709	7849	371	0
42257785	8740	8763	21.6	6.05	410	52	13063	5741	1604	510
42257468	15242	15290	21.5	7.85	1669	360	11340	7838	277	0
42018730	9322	9331	21.3	6.5	2210	0	11100	7910	80	18
42018952	10144	10154	20.9	8	1420	317	11200	7840	60	24
42004741	10672	10677	20.8	7.01	1303	58	11413	7793	150	7
42257467	12432	12498	20.6	5.6	140	10	12758	4798	2160	526
42016956	10553	10574	20.4	6.4	214	25	12300	6950	750	134
42019883	7310	7314	20.3	7.2	580	-3	11900	7314	486	29
42258090	10862	10936	20.1	7.93	1625	88	10745	7327	64	67
42257705	10850	10960	20.0	7.9	1352	280	10771	7403	113	22
42257706	10850	10860	19.9	7.9	1329	200	10771	7357	113	22
42005128	9529		19.9	7.7	1220	9770	2400	5890	600	61
42257801	10883	11404	19.6	8.4	1135	25	10824	7353	118	12
42016517	10860	10870	18.7	7.4	1415	12	10200	6800	180	79
42257480	11697	12511	18.7	6.84	1160	20	10360	6915	140	12
42257820	9392	9414	18.0	8.45	344	0	10693	6279	421	148
42018601	11052	11062	17.9	7.3	1520	35	9600	6560	172	34
42257755	8038	8053	17.6	7.65	490	360	9928	6612	81	17
42004345	10065	10067	17.6	7.8	878	7900	2900	5660	150	91
42257493	5747	8766	17.4	7.35	1368	-3	9431	6457	125	17
42258324	14153	14381	17.4	8.33	654	108	9940	6028	565	21
42257817	9360	9424	17.3	7.05	272	0	10315	6001	502	100
42257451	12150	12170	16.8	7.7	1460	0	8975	6149	153	0
42257465	12432		16.7	7.86	1340	235	8820	6050	230	15
42257757	8485	8493	16.6	6.8	223	82	9792	5740	287	142
42018600	10646	10682	16.5	7.1	1770	90	8500	6000	140	35
42252781	8232	8247	16.4	7.23	1079	242	8862	5986	180	40
42257466	12432		15.4	6.6	610	4	8883	5505	394	16
42258297	14974	14997	15.2	6.84	637	212	8520	5291	481	0
42009897	11090	11118	14.9	7.32	2000	69	7310	5398	61	12
42257509	9042	9060	14.5	7.93	1219	520	8463	5689	184	160
42018065	8967	8990	14.2		2315	30	6709	5075	14	3
42004759	6064	6081	14.1	7.38	1139	12	7666	4985	252	75
42257469	15245	15290	14.0	7.4	938	620	7219	4650	545	31
42006894	7102	7130	13.9	6.8	1390	1	7300	5170	54	16
42257683	9360	9396	12.9	5.7	200	0	7626	4632	280	36

Note: Italic values at the top of each column are averages of the individual well water analyses, USGS (2002).

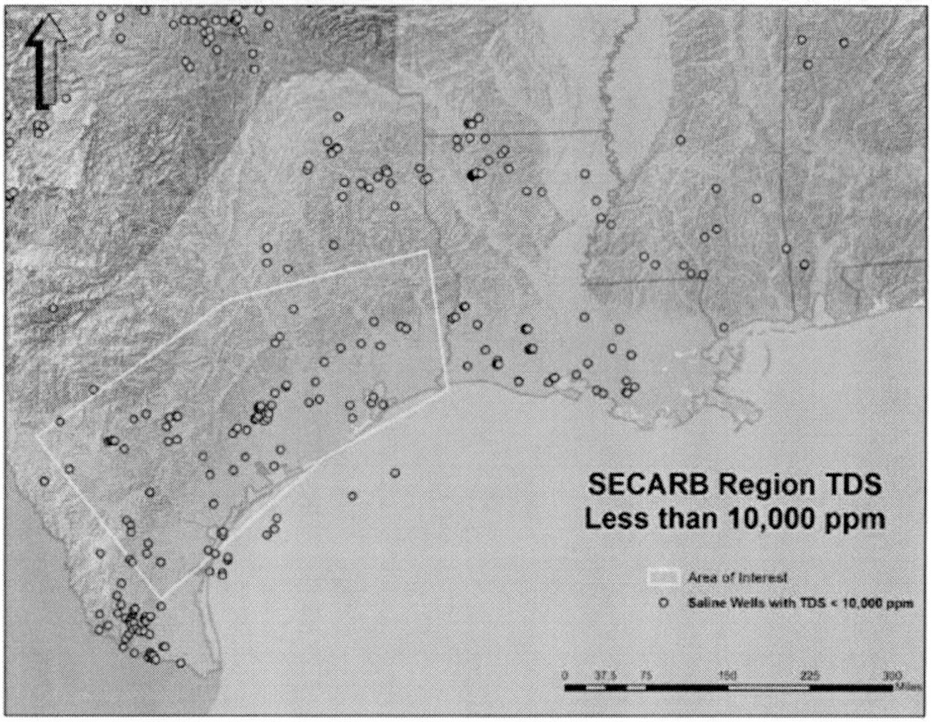

Figure A1a. NATCARB Wells in the SECARB Area with TDS Values below 10,000 ppm.

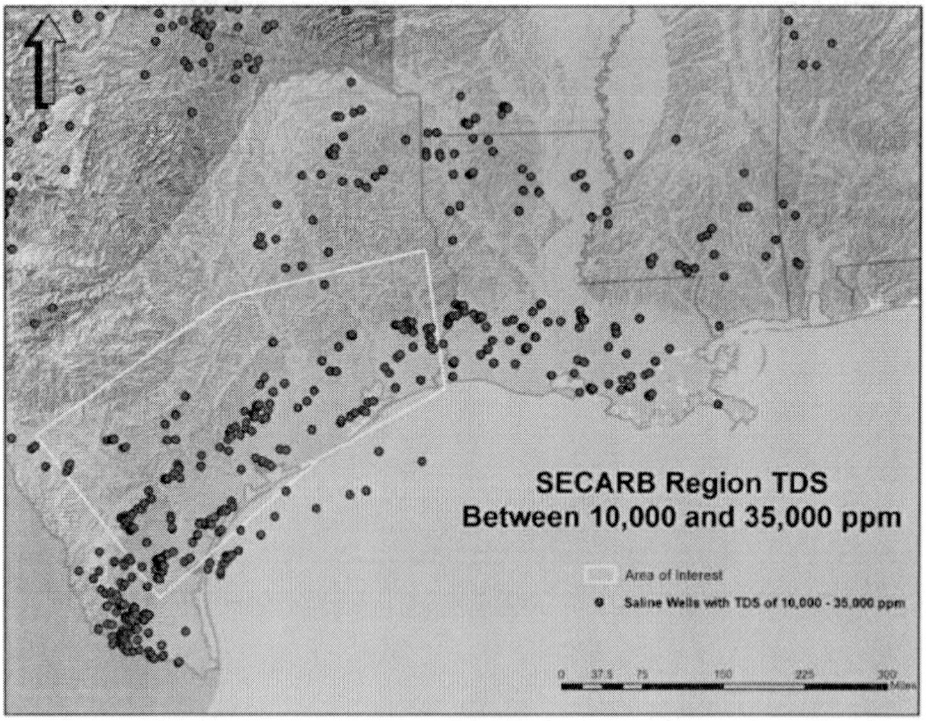

Figure A1b. NATCARB Wells in the SECARB Area with 10,000 – 35,000 ppm TDS Values.

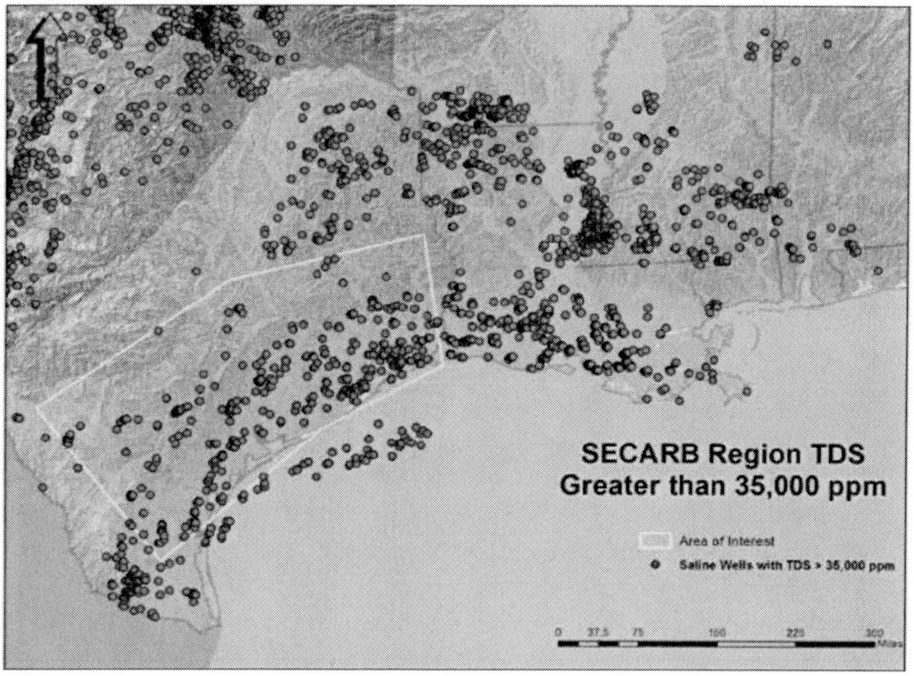

Figure A1c. NATCARB Wells in the SECARB Area with TDS Values Greater Than 35,000 ppm.

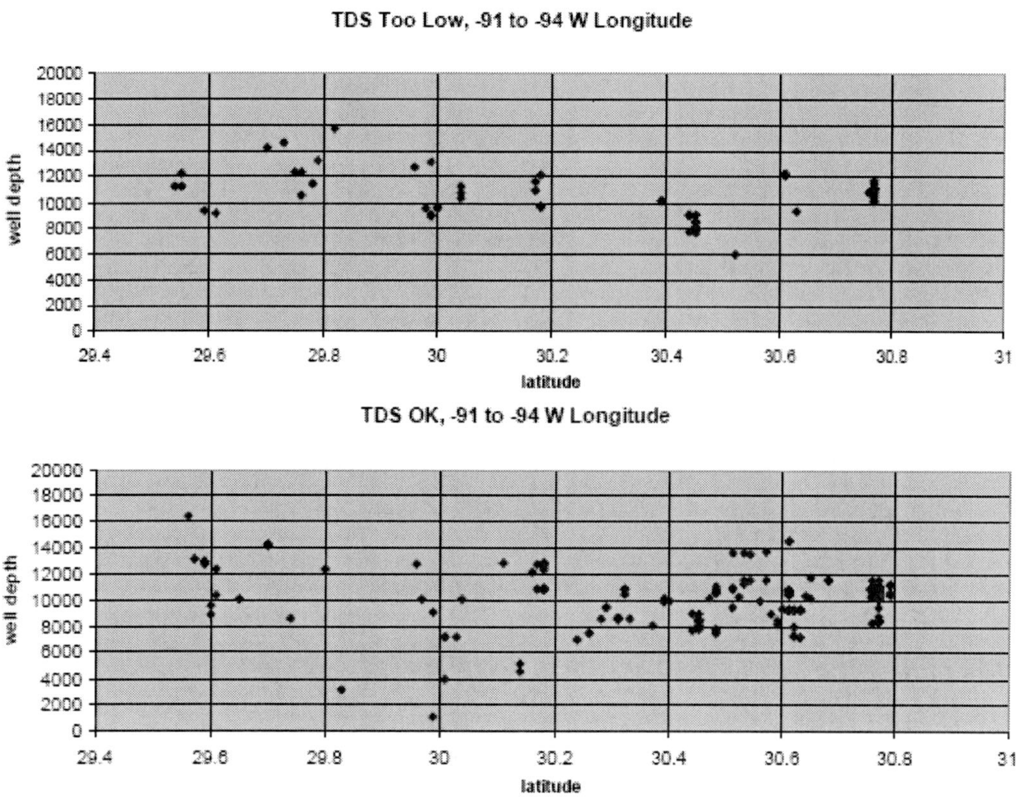

Figure A2. (Continued).

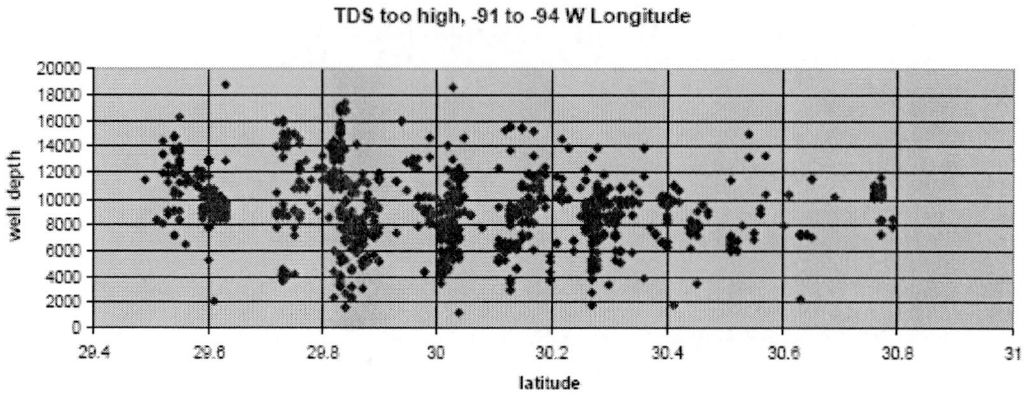

Figure A2 a,b,c. Vertical Section Illustrating Well Depth vs. Latitude for Low-, Medium-, and High-Salinity Fluid Wells in an East-West Swath Crossing Southern Louisiana, along with Well Depth in Feet.

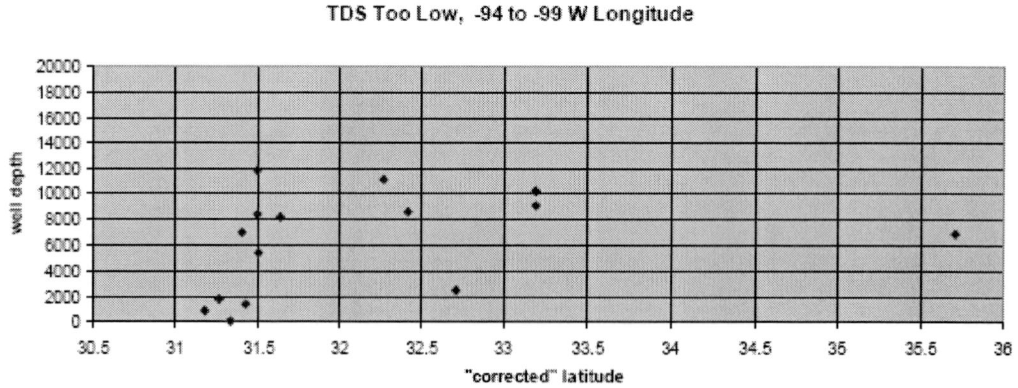

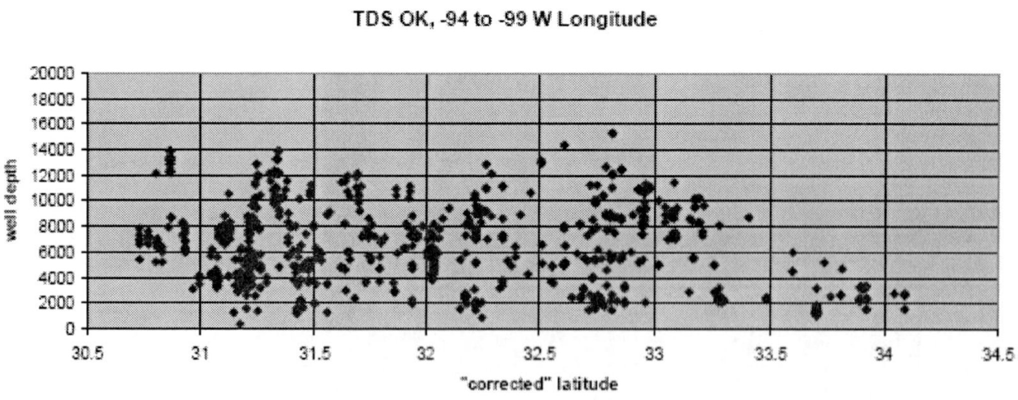

Figure A3. (Continued).

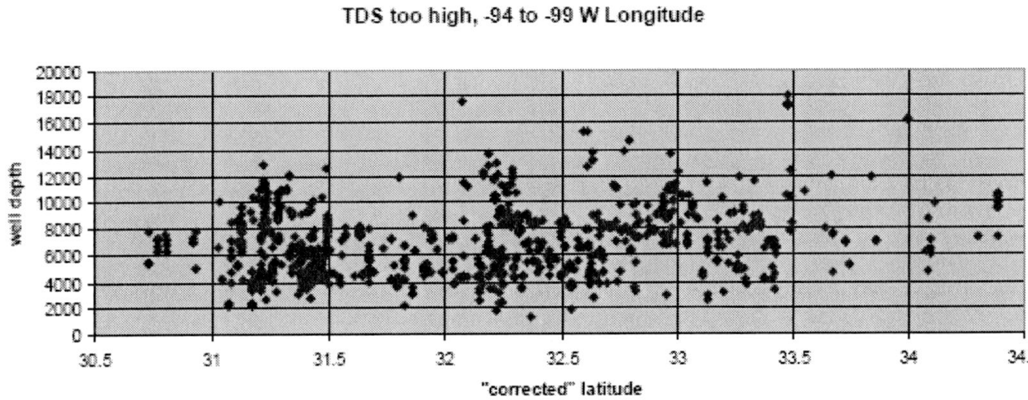

Figure A3a,b,c. Vertical Section Illustrating Well Depth & 'Corrected' Latitude for Low-, Medium-, and High-Salinity Fluid Wells in a Southwest-Northeast Trending Swath, Southern Texas, along with Well Depth in Feet.

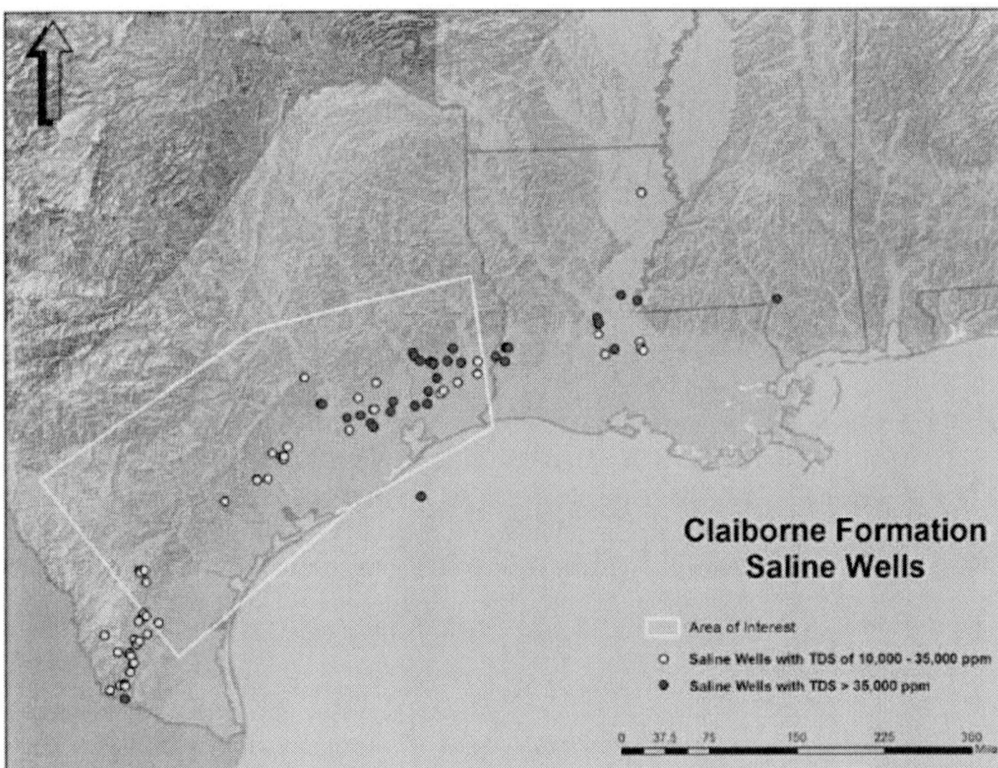

Figure A4. Distribution of High (Purple) and Mid-Salinity (Yellow) Wells, Claiborne Formation.

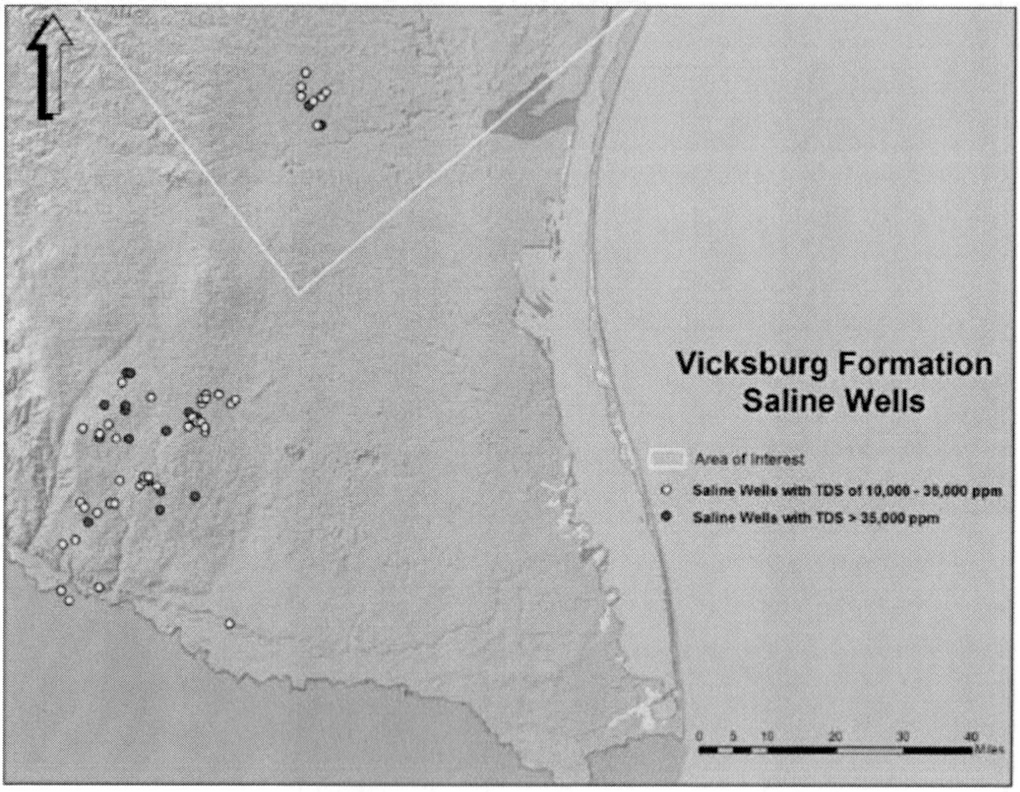

Figure A5. Distribution of High (Purple) and Mid-Salinity (Yellow) Wells, Vicksburg Formation.

Notes and Useful General Information Sources used to develop the extended Geoassessment presented in Appendix 1:

1. The USGS groundwater atlas: Referenced as USGS, 1999, U.S. Geological Survey GROUND WATER ATLAS of the UNITED STATES. The "home page" for the on-line version of the atlas can be found at http://pubs.usgs.gov/ha/ha730/gwa.html. However, while compiling this summary, numerous figures were extracted so, where applicable, the actual URL for that figure will appear with the figure caption.
2. For general USGS publications: http://pubs.er.usgs.gov/, and for the 2002 version of the USGS produced water database: http://energy.cr.usgs.gov/prov /prodwat/index.htm.
3. The NATCARB database contains a large compilation of well water chemistries from various formations; it can be accessed from http://drysdale.kgs.ku.edu/natcarb /eps/natcarb_alpha_content.cfm As of October 26, 2009.
4. Hovorka, S. D., M. L. Romero, A. G. Warne, W. A. Ambrose, T. A. Tremblay, R. H. Treviño, and D. Sasson, 2000, "*Sequestration of Greenhouse Gases in Brine Formations*"; Texas Bureau of Economic Geology: http://www.beg.utexas.edu /environqlty/co2seq/dispslsaln.htm. Then go to "Examine Data" and to "Data Matrix."
   Alternatively, regional brine chemistry summaries can be accessed directly through: http://www.beg.utexas.edu/environqlty/co2seq/co2data.htm.

5. Energy Citations Database: http://www.osti.gov/energycitations/
6. Texas formation water chlorides database: http://gswindell.com/texasrw.htm#accesstxrw
7. USGS Produced Water Database, released 2002: http://energy.cr.usgs.gov/prov/prodwat/index.htm
8. Georgia Geological Survey, 1999.
9. Geological Survey of Alabama, 2006.

APPENDIX B

Analytical Model

This section addresses the question of supercritical CO_2 migration in a gently sloping confined saline formation with no structural closure. This situation could arise, for example, following injection of CO_2 into Gulf Coast saline formations. This may be a concern because the buoyant CO_2 could migrate toward the earth's surface beneath a caprock and intersect the surface where the caprock/reservoir contacts outcrops. Use of a reservoir simulator (e.g., TOUGH2) to address this question was completed; however, the long transport distances in question require long numerical times for adequate resolution. A quicker, simpler way to constrain migration distances is to use a simplified numerical approach as was examined recently in Hesse et al. (2008; 2009).

Hesse et al. (2008; 2009) derive a simple model for the temporal and spatial evolution of the (assumed) sharp interface between injected CO_2 and brine, and account for residual trapping of CO_2 in the wake of the gravity current (buoyant CO_2 moving along the saline formation/caprock interface). These authors consider the flow of supercritical CO_2 with density ñ and viscosity μ_c, within a brine reservoir with brine density and viscosity of ñ+Δñ and μ_b respectively. The buoyant CO_2 flows within a sloping infinite saline formation of constant thickness H and dip angle è. They neglect velocities in the z (vertical) direction and assume that pressure in each fluid type is hydrostatic. They also assume that capillary equilibrium is maintained. These are somewhat standard assumptions in studies examining sweep efficiency and fluid displacement made by petroleum engineers (see Hesse et al., 2008, for references). Other simplifying assumptions include constant and homogeneous saline formation flow properties and porosity.

Using conservation of mass principles, these authors derive the following model for the post-injection evolution of the CO_2 plume thickness h (which also locates the CO_2 brine interface in the 2-D aquifer studied).

$$\frac{\partial h}{\partial t} = \kappa \cdot \frac{\partial}{\partial x}\left[\frac{h(H-h)}{h(M-1)+H}(-\sin\theta + \frac{\partial h}{\partial x}\cos\theta)\right] \tag{1}$$

Where M is the mobility ratio (measuring the contrast in the 2-phase mobilities across the advancing or contracting interface;

$M = \frac{k_{rc}}{\mu_c}\frac{\mu_b}{k_{rb}}$ for $k_{ri} =$ relative permeability of phase i) and the CO_2 conductivity κ is given by

$$\kappa_1 = \frac{k\lambda_c \Delta\rho g}{\phi(1 - S_{br} - S_{cr})} \; for \; \frac{\partial h}{\partial t} < 0 \quad and \quad \kappa_0 = \frac{k\lambda_c}{\phi(1 - S_{br})} \; for \; \frac{\partial h}{\partial t} > 0 \tag{2}$$

Here the residual brine saturation S_{br} and the residual CO_2 saturation S_{cr} are assumed constant.

Letting

$$\eta = \frac{h}{H}, \; \xi = \frac{x}{L}, \; \sigma = \frac{\kappa}{\kappa_1}, \; and \; \tau = \frac{t}{L(\kappa_1 \sin\theta)^{-1}}$$

and substituting into (1) yields the following dimensionless equation (L is the width of the initial injected plume):

$$\frac{\partial \eta}{\partial \tau} = -\sigma \frac{\partial f}{\partial \eta}\frac{\partial \eta}{\partial \xi} + \frac{\sigma}{P_e}\left[\frac{\partial f}{\partial \eta}\left(\frac{\partial \eta}{\partial \xi}\right)^2 + f\frac{\partial^2 \eta}{\partial \xi^2}\right] \tag{3}$$

with σ expressing the step function in equation (2) as

$$\sigma = 1 \; when \frac{\partial \eta}{\partial t} < 0 \; (thinning\; plume) \; and = 1 - \varepsilon \; when \frac{\partial \eta}{\partial t} > 0 \; (thickening \; plume) \tag{4}$$

A numerical solution to (3), (4), and (5) is conveniently found using the NDSolve routine in MATHEMATICA (Wolfram, 2003). As an example, we use the discussion and parameters in Hesse et al. (2008), relevant to the storage of CO_2 in the Carrizo-Wilcox saline formation (Figure B1) derived from the study by Nicot (2008). Parameters are as follows: H = 200 km, k = 500 mD, φ = 0.15, average injection depth equals 2.7 km and so $\Delta\rho = 300$ kg/m^3, dip angle $\theta = 1.5^{\circ}$, $S_{cr}=S_{br}=0.2$, $\mu_b/\mu_c=10$, $k_{rc}=0.2$, and $k_{rb}=1$. At the end of 50 years of injection, the average horizontal size of the plume L is given as 10 km, and the initial aspect ratio of the plume A = L/H = 50. This yields parameters for (3) of $P_e = 1.4$, $\varepsilon = 0.25$ and $M = 5$.

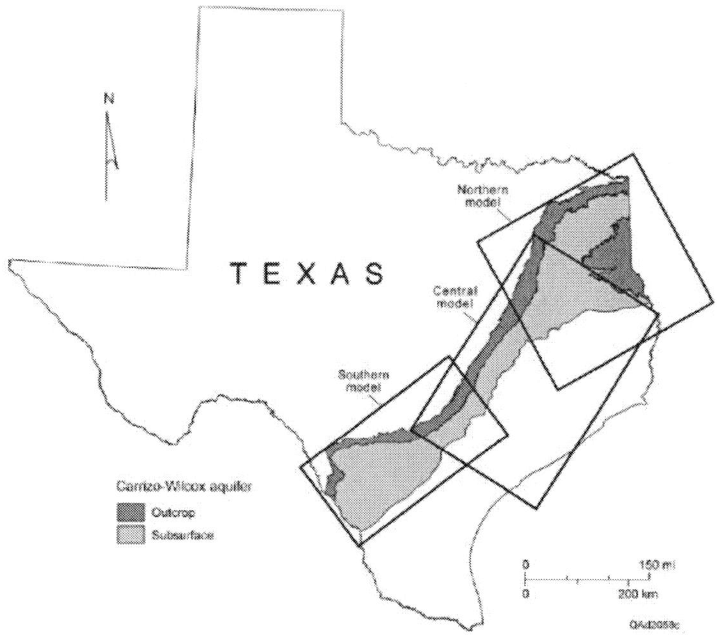

From Dutton et al., 2003.

Figure B1. Outcrop and Subsurface Extent of Carrizo-Wilcox saline formation.

With these parameters, the solution to (3-5) is shown graphically below for dimensionless times $\tau = 0$, 15, 75, 100, and 150 (updip is in the direction of increasing ζ). Here we plot only the remaining plume (i.e., we do not plot the residual CO_2 trapped behind the gravity wave).

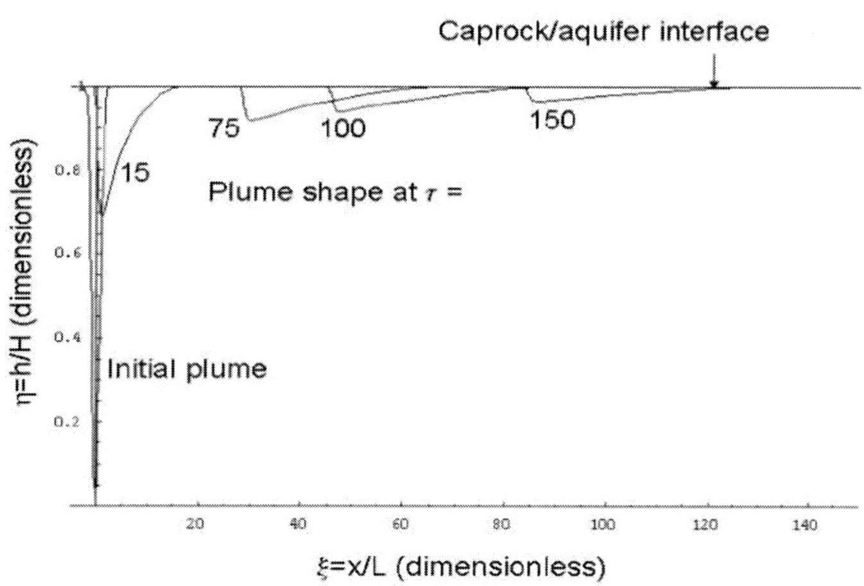

Figure B2. Temporal and Spatial Evolution of Plume of Mobile Supercritical CO2, from Numerical Solution of Model in Equations (3, 4, and 5).

The full time evolution of the morphology of the front is depicted in Figure B3:

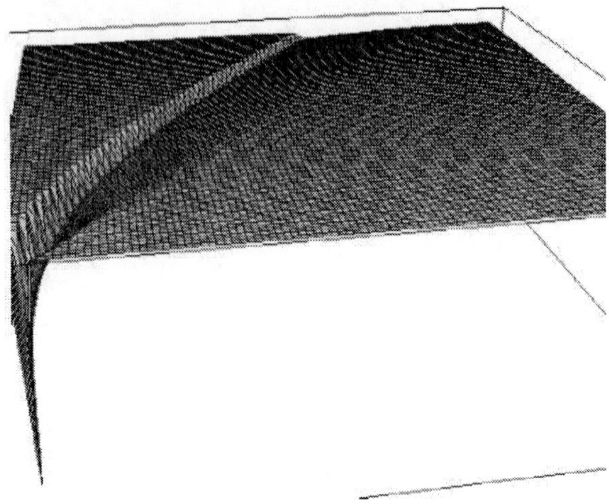

Figure B3. 3-D View of Plume Shape Evolution.

The values of t = 15, 75, 100, and 150 have maximum î distances roughly equal to 15, 65, 85, and 125 respectively. For L = 10 km, k = 1.93e-5 m/s, yielding a characteristic time t_a ~ 820 years. In dimensional form, these t and î values correspond to the following:

τ max	ζ max	t_{max}(thousand years)	X_{max} (km)
15	15	12	150
75	65	61	750
100	85	82	1000
150	125	123	1500

With an average distance from subsurface to outcrop in the Carrizo-Wilcox saline formation of, say, 100 km (Figure B1), these results suggest that in no way will all CO_2 be trapped as an immobile residual phase; that is, CO_2 will remain mobile to the surface. In addition, it would take on the order of 9,000 years to reach the surface if this model and the above parameters are at all representative. Note that in addition to all the above-mentioned assumptions and simplifications inherent in this model, there is no account of the phase change that would encounter the CO_2 (i.e., from supercritical to gas phases) as it rises through the saline formation.

This analysis could easily be adapted to address similar question of CO_2 migration and extent in other Gulf Coast saline formations. Note for simplicity, we used an initial plume shape equivalent to a Gaussian distribution and that other initial plume forms could be used.

Figure B4 shows a geologic framework model of the layered clastic system prepared from formation tops extracted from the Dutton et al. (2003) and Kelley et al. (2004) GAMs and kriged using C-Tech's MVS software. This framework model was used to construct the TOUGH2 grid as shown in Figure B5. The coarse grid has been refined to allow for good numerical resolution around injection and abstraction wells. Formation permeability and heterogeneity therein have been mapped onto the model grid using the distribution of hydraulic conductivity derived from the GAM. This approximates the down-dip fining of the regressive sand-silt-shale sequences, locations of fault systems, and the sandier portions of fluvial-deltaic sequences making up these formations (which may act as CO_2 conduits).

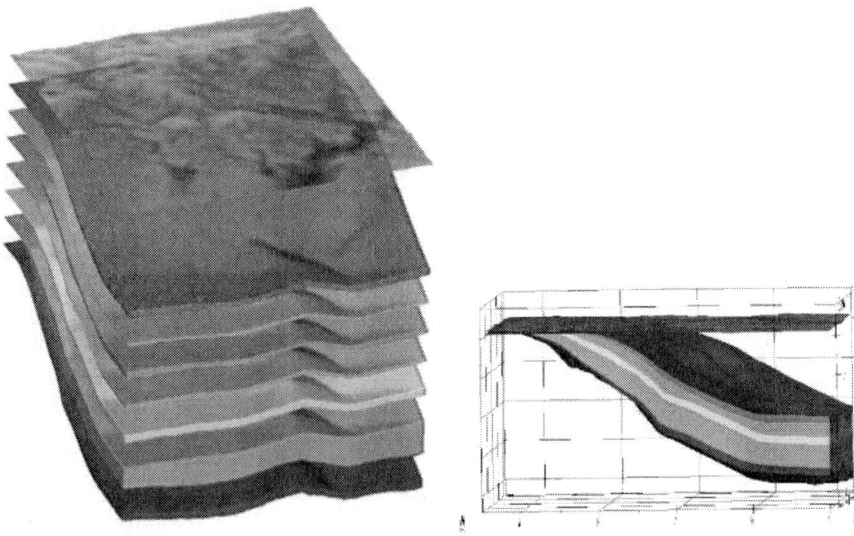

Figure B4. Geologic Framework Model for Carrizo-Wilcox System. This refers to the bold rectangle in Figure 3-1, created using formation tops and thicknesses from the Texas Water Development Board (TWDB) Groundwater Availability Model. Both left and right figures assume a 20x vertical exaggeration.

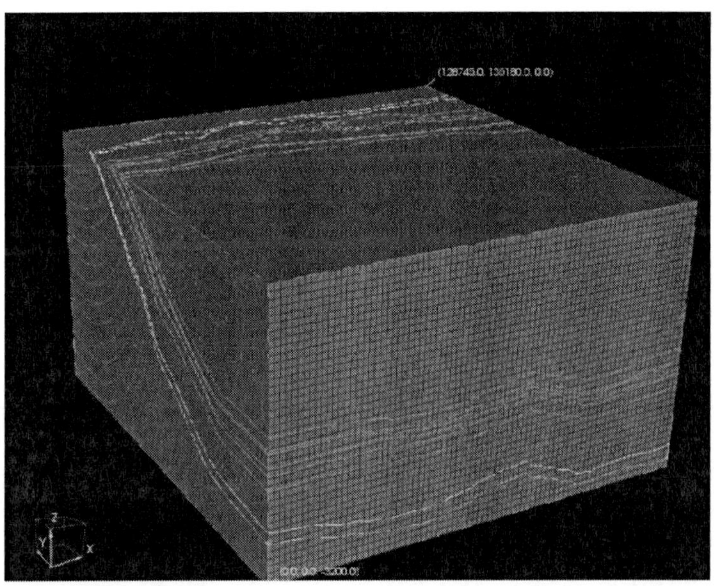

Figure B5. TOUGH2 Simulation Domain. Coarse gridding at approximate size of 1.5 km in x and y and 80 m in z, with 20x vertical exaggeration. The Simsboro Formation shown in yellow and the overlying Calvert Bluff and Carrizo Formations are suggested targets for CO_2 storage.

In the Figure B4, the exploded view (left) and cross section (right) assume 20x vertical exaggeration. The red layer is the Simsboro Formation, orange is the Calvert Bluff (both of the Wilcox Group) yellow-green is the Carrizo Formation, and the remaining overlying layers are the Reklaw, the Queen City, Weches, and Sparta Formations in order of decreasing depth. The top surface shown is derived from a USGS digital elevation model for the region. The cross section on the right is a little south of the cross section line depicted by Nicot (2008).[5]

Boundary conditions for the TOUGH2 model assume closed flow boundaries on the sides and at the Wilcox growth fault boundary and are open to flow at the top. Later versions of this model will explore allowing heavy CO_2-laden aqueous solutions to sink by gravity into the lower Hooper Formation of the Wilcox Group (currently disabled).

Simulations

Preliminary TOUGH2 runs have been completed on a simplified version of the grid because of size constraints by TOUGH2 gridblock naming conventions, which limit the total number of gridblocks to 99,999 cells. For these runs, we use a grid system 20x42x80 in x, y, and z. Our region is 0 to 128747 m in x, 0 to 135185 m in y, and -3540 to 0.0 in z (for simplicity, we have taken the top of our domain to be at sea level). This produces a grid block size of 6437.4 m in x, 3218.7 m in y, and 40.25 m in z. This is not as good as the horizontal resolution of the Carrizo Wilcox GAM (1 square mile) but improves on the vertical resolution (the GAMs use 1 grid block per layer for 6 layers). An injection well was completed over a 100 m interval at {x,y,z}= {5.5255e4, 2.7514e4, -2.85412e3} where the z value denotes bottom hole. At this depth, all CO2 is injected as a dense supercritical fluid, but as density decreases as a CO2 plume rises as a result of temperature and pressure decreases, the volume occupied by the same mass of CO2 increases.

This coarse grid is shown in Figure B6. Cells down-dip of the Wilcox Growth Fault region are disabled as are cells beneath the target Simsboro Formation. The growth-fault zone is thus taken as a hydrological boundary as was done for the Carrizo-Wilcox GAM. A pressure gradient of 0.01 MPa/m, with surface pressure of 0.1 MPa, was applied as an initial condition. An initial temperature distribution was given by a surface temperature of 15 °C and a vertical gradient of 25 °C/km. Finally, a salinity gradient equaling 50,000 ppm per 3,000 m was applied initially, with a surface TDS of 500 ppm (assuming all dissolved solids as NaCl), was applied to the model. A linear increase of salinity with depth is seen in virtually all Gulf Coast formations (Dickey, 1969); our range is consistent with observed salinities in the normally-pressured portion of the Wilcox Formation of interest here (Figure 3-1) but is less than is observed farther north in the Wilcox off the coast of Texas and Louisiana where Wilcox groundwater salinities show impact by presence of salt domes.

A series of CO_2 injection simulations was undertaken to check sensitivities of unknown parameters. Injection of CO_2 was maintained constant at 1 MTonnes/year. The most uncertain of the parameter set for the Carrizo-Wilcox GAM and required by the TOUGH2 runs is vertical permeability (Dutton et al., 2004). Anisotropy in hydraulic conductivities (denoted as a ratio of vertical to horizontal conductivities = Kv/Kh) ranges from unity to 10^{-4}. Although heterogeneity in hydraulic conductivity was constrained by the calibrated MODFLOW models in the GAM, we assume average permeabilities for each of the formations (calculated from the GAM—we will investigate the influence of permeability heterogeneity as part of our ongoing study—this will be reported elsewhere) and examine variations in the Kv/Kh values. Figures B7 to B10, corresponding to a Kv/Kh value of 1.0, 0.1, 0.01, 0.001 respectively, show a range of behaviors from complete vertical leakage at Kv/Kh = 1.0, to nearly complete sealing and updip migration of the CO_2 plume through mostly the Carrizo Formation at Kv/Kh = 0.01. Reasonable values of Kv/Kh in the Reklaw Formation range from 0.01 to 0.001 (Dutton et al., 2003; Kelley et al., 2004), so it seems reasonable to expect that while most CO_2 will migrate updip, a substantial amount will leak into the Reklaw and above formations.

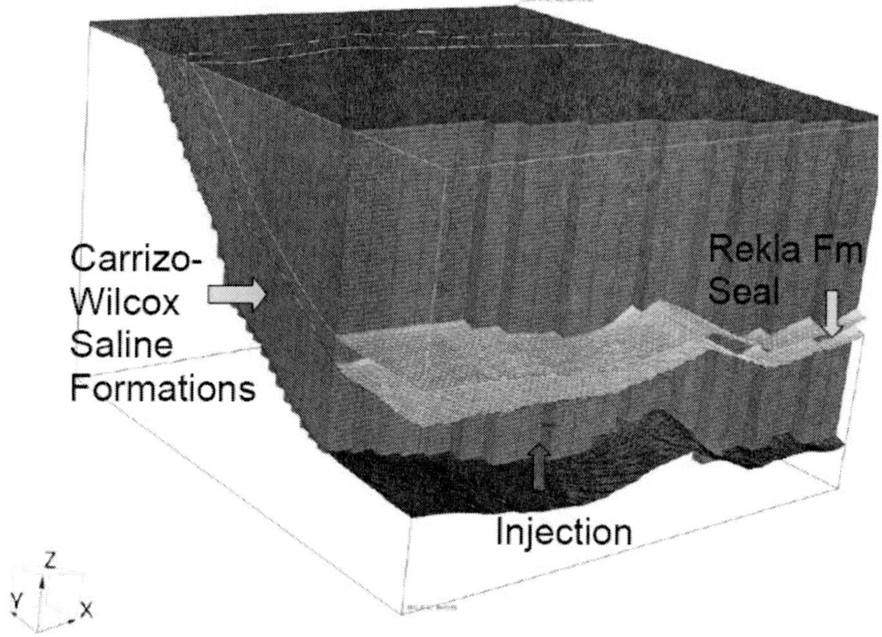

Figure B6. Updip Limit of Wilcox Growth Fault System (which acts as the eastern hydrologic boundary, and coarsened grid for regional TOUGH2 simulations). Yellow surfaces mark the upper and lower extent of the Eocene Reklaw Formation, which acts as a regional hydrologic seal. Injection zone in the Simsboro Formation of the Wilcox Group is shown by red arrow. Figure has 20x vertical exaggeration.

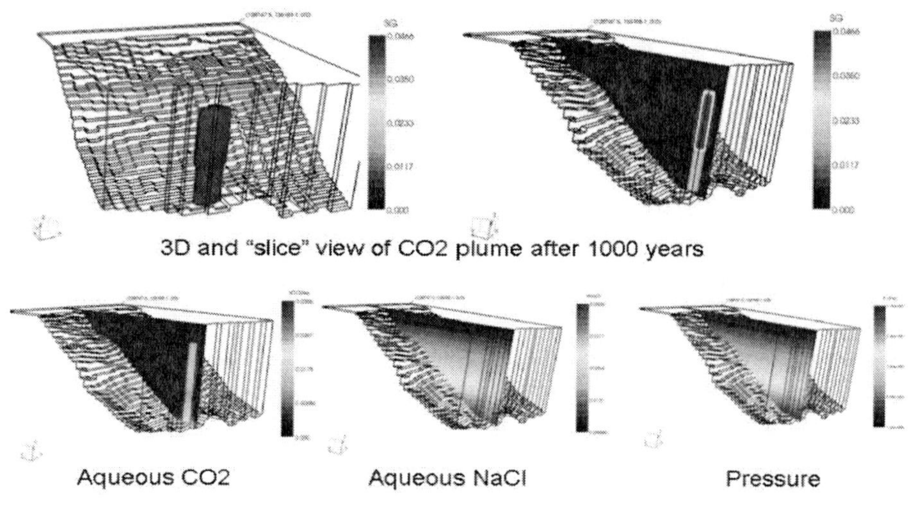

Figure B7. CO_2 Saturation, Aqueous CO_2, Aqueous NaCl, and Pressure (in Pa) in an Injection Scenario where CO_2 Is Injected at a Rate of 1 MTonnes/Year. The Reklaw seal in this case is taken to have homogeneous isotropic permeability equal to 3.0e-15 m^2. In this case, the CO_2 plume ponds beneath the Rekla, but dominantly migrates vertically displacing aqueous NaCl and pressure slightly upwards into the overlying strata. Figure shows 20x vertical exaggeration.

In Figure B7, with isotropic and homogeneous permeability in the Reklaw seal (which is two orders of magnitude less than the average permeability in the underlying Carrizo-Wilcox System), CO2 is seen to migrate vertically through the overlying strata and not confined beneath the Reklaw seal. Breakthrough to the surface occurs in ~1,500 years. There is a slight perturbation of salinity and fluid pressure above the rising plume in the overlying formations.

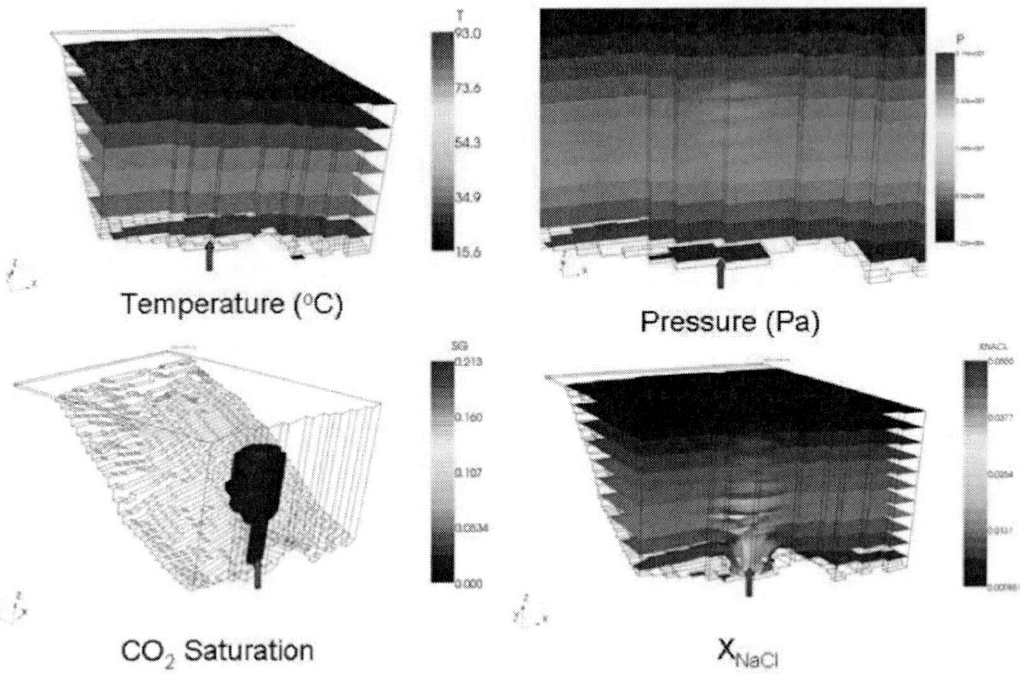

Figure B8. Temperature, Pressure, Gas Saturation, and Mass Fraction of Aqueous NaCl after 4,770 Years of CO2 injection at Constant Rate of 1 MTonnes/Year with Kv/Kh = 0.1. Figure shows 20x vertical exaggeration.

In Figure B8, showing a similar simulation but with a slight Kv/Kh ratio of 0.1, there is more ponding of the CO2 plume, but movement is still dominantly vertical, and breakthrough to the surface occurs after ~5,000 years. Pressure is slightly elevated above the plume in the shallow portions of the model. The CO2 "ponds" beneath the Reklaw seal and spreads, but substantial vertical leakage is observed. Salinity is seen to "dry-out" from the pore spaces occupied by the CO2 plume, and some amount of more saline water is seen to intrude into the overlying formations.

In Figure B9, with Kv/Kh of 0.01, the Reklaw here acts as a decent seal and CO2 is confined to updip migration within the Carrizo Formation. Decreasing the Kv/Kh to 0.001 (Figure B8) ensures a reasonable capillary seal to vertical migration, and the CO2 plume migrates just beneath the Reklaw within the Carrizo Formation. The migrating "tongue"-shaped plume reaches the location of existing groundwater wells in the Carrizo (shown in yellow in Figure 3-1) after approximately 5,000 years. Enriched aqueous CO2-bearing waters will reach the Carrizo wells sometime just prior to arrival of gas-bearing waters. It is important to note also that of the 5,000 MTonnes (5e12 kg) injected in this model, about half (2,512 MTonnes) exists as a supercritical phase, and half (2489 MTonnes) is dissolved in the surrounding aqueous phase.

Figure B9. CO_2 Saturation, Aqueous CO_2, Aqueous NaCl, and Pressure (in Pa) in an Injection Scenario Where CO_2 Is Injected at a Rate of 1 MTonnes/Year. The Rekla seal in this case is taken to have a Kv/Kh ratio of 0.01, and as a result, the CO_2 is retained beneath the Reklaw and migrates up dip through the Carrizo formation. There is little displacement of the pressure contours, and a slight updip migration of salinity, after 1000 years. Figure shows 20x vertical exaggeration.

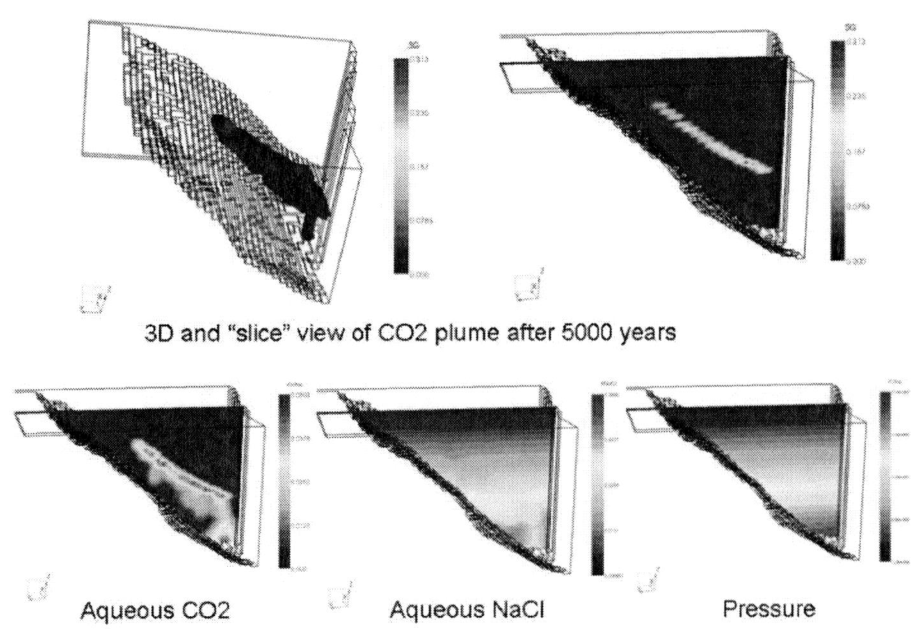

Figure B10. CO_2 Saturation, Aqueous CO_2, Aqueous NaCl, and Pressure (in Pa) in an Injection Scenario Where CO_2 Is Injected at a Rate of 1 MTonnes/Year. (The Reklaw seal in this case is taken to have a Kv/Kh ratio of 0.001, so the CO_2 is retained beneath the Reklaw and migrates updip through the Carrizo Formation.) After 5,000 years, there is a slight updip migration in salinity and a very slight pressure disturbance propagating updip. Figure shows 20x vertical exaggeration.

Future work could include the effects of including permeability heterogeneity, particularly the updip increase in vertical permeability, taken from the Carrizo-Wilcox and Queen City-Sparta TWDB GAMs (Dutton et al., 2003; Kelley et al., 2004). Also, one could include additional injection wells within the simulation domain to accommodate 90% of the 11.9 Mtonnes of CO_2 output by the nearby Fayette Power Project plant and associated water-withdrawal wells for power plant usage.

End Notes

[1] Sprinkle (1989) illustrate this in Figure 27 of their work.
[2] USGS (1999) illustrates this in Figure 51 of their work.
[3] Smyth et al. (2007) illustrates this in Figures 2, 3 and 4 of their work.
[4] HERO™ is a registered trademark of Debasish Mukhopadhyay.
[5] Nicot (2008) illustrate this in Figure 1 of his work.

INDEX

D

E

F

G

H

I

J

K

L

M

N

S

T